BIOLOGICAL SPECIFICITY
AND GROWTH

THE TWELFTH SYMPOSIUM OF
THE SOCIETY FOR THE STUDY OF
DEVELOPMENT AND GROWTH

Biological Specificity
and Growth

A. M. SCHECHTMAN · M. W. WOERDEMAN

M. R. IRWIN · S. C. SHEN · R. BROWN AND E. ROBINSON

J. R. RAPER · J. A. HARRISON · W. H. TALIAFERRO

H. S. N. GREENE · P. WEISS · R. BRIGGS AND T. J. KING

EDITED BY ELMER G. BUTLER

PRINCETON, NEW JERSEY

PRINCETON UNIVERSITY PRESS

1955

FOREWORD

THE Society for the Study of Development and Growth annually conducts a symposium which brings together for several days a group of active investigators to discuss important problems in the field of growth. Participants represent many scientific disciplines, including not only biology but also the related fields of chemistry, physics, and the medical sciences. Guest speakers present prepared papers in which they report recent research in their laboratories, review progress in the general area with which they are concerned, and often point the way to fresh interpretations and new paths of investigation. Problems of both abnormal and normal growth receive attention. So far as possible, the symposia are conducted in an informal manner; vigorous discussion in the lecture hall is often continued at the dinner table and later in informal groups on the grass under the trees. The prepared papers, frequently revised in the light of discussions at the time of the symposium, become the chapters of the symposium volume.

The topic chosen for the Twelfth Growth Symposium was *Biological Specificity and Growth*. In developmental biology the phenomenon of specificity occupies a prominent position. When it can be determined that interactions among components of a biological system are such that specific activities occur, or when a particular stimulus, internal or external, evokes a specific response, it becomes of considerable importance to discover the basis for the selectivity and specificity concerned. It is with an examination and evaluation of the role of specificity in growth and differentiation that this volume deals. The participants in the symposium have approached the problem at different levels of organization and by the use of a variety of biological materials and techniques. Taken together the chapters of this volume represent an assessment of our present knowledge of the subject.

This symposium was planned jointly by the Society for the Study of Development and Growth and the Committee on Developmental Biology of the National Research Council. It was held at the University of New Hampshire, Durham, June 14-22, 1953. Grateful acknowledgment is made to the members of the biological faculty and to the administration of the university for their hospitality and for the excellent facilities they so generously provided. Financial support which made the symposium possible was received from the American Cancer Society, acting through the Committee on Growth of the National Research Council; the National Cancer Institute of the National Institutes of Health; the National

Science Foundation; and the Committee on Developmental Biology of the National Research Council.

The appearance of growth symposia in book form began with the Eleventh Symposium on the *Dynamics of Growth Processes,* published by the Princeton University Press in 1954. Earlier symposia appeared as supplements to the journal *Growth.* Limited quantities of these supplements are still available from the editors of that periodical.

All of the initial editorial work on this volume was carried out by Dr. David R. Goddard, University of Pennsylvania. Unfortunately, because of the press of other duties he was prevented from continuing with the editorship. Dr. John T. Bonner, Princeton University, has rendered considerable assistance at various stages in the preparation of this book. Thanks are also due to Miss Suzanne H. Eldredge, Princeton University, for continued assistance with manuscript and proof. All those concerned with the preparation of this volume acknowledge with gratitude the cooperation of Mr. Herbert S. Bailey, Jr., of the Princeton University Press.

ELMER G. BUTLER

Princeton University
May 1954

CONTENTS

BIOLOGICAL SPECIFICITY
AND GROWTH

I. ONTOGENY OF THE BLOOD AND RELATED ANTIGENS AND THEIR SIGNIFICANCE FOR THE THEORY OF DIFFERENTIATION

BY A. M. SCHECHTMAN[1]

I. THE NATURE OF IMMUNO-EMBRYOLOGICAL INVESTIGATION

ALL MODERN HYPOTHESES concerning differentiation mechanisms are concerned in the last analysis with molecular specificity. It is interesting to note that most of our recent hypotheses concerning the creation of new molecular specificities stem from work done in other fields of biology which just a few years ago had no apparent relationship to embryonic development. These hypotheses suggest possible means by which molecular specificities may be altered, abolished, or created. Some, like the concept of antibody formation applied to development, have already stimulated experimentation. Others, like the concept of adaptive enzymes, have hardly advanced beyond the "talking stage" insofar as embryonic differentiation is concerned. The speculations of Weiss (1947), Tyler (1947), Brachet (1949), Sonneborn (1947), Schultz (1952), and others have helped greatly to clarify the essence of the problems of differentiation as well as to suggest processes which possibly go on within the embryo. The purpose and design of experimental work for many years to come will doubtlessly spring from these speculations.

The patterns of molecular specificities, established by means as yet highly hypothetical, constitute the groundwork of differentiation and are expressed according to the means of analysis at our disposal: as morphological differentiation, biochemical change, specific physiological functions, embryonic competences, etc. With such assumptions as to the fundamental nature of differentiation, it is little wonder that immunological phenomena have become of greater interest to the embryologist as well as to others concerned with the creation of new features by living materials. A process which permits the production of apparently endless numbers of specific molecules in the form of antibodies, and these specificities resident in protein molecules which may be chemically identical, would seem to have possibilities leading toward the understanding of causal factors in the production of new features in the developing organism.

[1] Department of Zoology, University of California, Los Angeles.

However, it must be said that during the past fifty years or so in which it has been applied to developmental problems, immunology has been used largely as a tool for the investigation of chemical epigenesis rather than the causes of epigenesis. In this respect it has been used like the microscope, which revealed the details of morphological epigenesis. Immuno-embryology has shown little of the approach characteristic of *Entwicklungsmechanik,* which deliberately seeks causal factors in epigenesis. The time of appearance or disappearance of antigens and the occurrence of certain antigens throughout the life history of the organism are descriptive chemical embryology and are equivalent to studies on the appearance or disappearance of morphological structures such as the notochord or pronephros. Of themselves they suggest, but they do little to elucidate the mechanisms by which specificity is established. Such studies certainly give us important correlations: an antigenic change precedes or is concomitant with a morphological change. I have no intention of depreciating the significance of these studies. The application of immunology to chemical epigenesis is a vital part of embryological investigation because (1) it demonstrates an order of chemical specificities which thus far are not detectable by any other methods and (2) it brings us closer to differentiation on the molecular level. A start has been made in the application of immunology to causal aspects of development in such experiments as those of Weiss (reviewed, 1950) and Ebert (1950). However, it is important to realize clearly the nature of the results thus far obtained by this method as it has been applied in experimental work. Only then can we apply immunological concepts and methods toward causal analysis of development.

II. THE SIGNIFICANCE OF MACROMOLECULAR TRANSFER

The main objectives of the present discussion are to show (1) that molecular epigenesis occurs during development and (2) that the new macromolecules which appear are not necessarily the products of the embryo's own synthesis. The specific data cited are largely derived from studies of the embryonic blood and related antigens. These provide the clearest though not the only evidence that new molecules like new morphological structures make their appearance during the course of development. It is equally clear that the *de novo* appearance of antigens or other molecules tends to create a false impression of the synthesizing capacities of the embryonic organism. The embryo is definitely deficient in some types of molecular synthesis and is dependent upon presynthesized molecules supplied normally by the maternal body. Of course it is obvious

[4]

that in one way or another the organic materials required by the early embryo are of maternal origin. But it has been assumed generally that the materials transmitted are of a rather simple nature, the kind that readily pass through the plasma membrane of most adult cells. It has been assumed further that such small molecules are nothing more than simple structural units or sources of energy which provide some of the "conditions of life" that permit the embryo to elaborate its own specific macromolecules. On this basis the mechanisms of differentiation reside wholly in the embryo. The situation assumes a totally different aspect when we realize that *the embryo admits complex macromolecules and that such admission occurs prior to and during the time when the basic differentiations of the vertebrate body are established.* We have, therefore, the essential relationships observed in certain microorganisms in which a macromolecular constituent, apparently desoxyribose nucleic acid, induces a permanent alteration in the synthesizing capacities of the organisms (McCarty and Avery, 1946, on *Pneumococcus;* Boivin et al., 1945, on *E. coli*). However, the great distinction between the status of bacterial transformation and embryonic differentiation lies in the fact that there is good evidence for a causal relationship between the macromolecule and transformation, whereas the uptake of macromolecules and embryonic differentiation have not yet been related causally.

In the following we shall refer to macromolecular substances elaborated by the embryo as autosynthetic and to those produced by the adult organism and transferred to the embryo as heterosynthetic.

III. EPIGENESIS OF RED BLOOD CELL ANTIGENS

It is a curious fact that embryologists are not inclined to attribute causal differentiative roles to the antigens of the red blood corpuscles. On the other hand it has been assumed that antigens appearing, for example, during the development of nervous, spleen, or cardiac tissue may have some causal role in the differentiation of these tissues. The reasons for this discrimination seem to be that blood cells differentiate whether or not they contain certain specific antigens; in addition much of the available evidence has indicated that the antigens are present from the earliest stages studied and therefore must be independent of morphological or physiological differentiation. Thus the A and B antigens were reported in the 37-day human fetus (Kemp, 1930), the Rh antigens in the fetus weighing only 8 grams (see Levene, 1948). Bornstein and Israel (1942) found agglutinogens A,B,M,N, and Rh positive in 7-21 cm. human fetuses, and Keeler and Castle (1933, 1934) found two antigens (H_1 and H_2) in the nucleated red cells of the 4 mm. rabbit embryo.

However, there are indications that certain of the human antigens are not static but undergo change in development. Thus the A and B antigens seem to be "incompletely developed" in the red cells at birth (Wiener, 1946). If we are looking for correlations we have one in this instance, for the maturation of the A and B antigens occurs concomitantly with the gradual disappearance of fetal hemoglobin and the rise of adult hemoglobin. The Lewis factor (Le^a) also apparently differs in the infant and adult. Antisera react with the adult factor only if it is in homozygous cells whereas the factor is detectable in both homozygous and heterozygous infants (Andresen, 1947).

Epigenicity is more clearly discernible in certain blood cell antigens of birds. Certain chicken red cell antigens are present in very early development whereas others do not appear until after hatching (Briles, McGibbon, and Irwin, 1948). Six alleles of the B series of antigens occur in the 3-day embryo. In the D series one factor is found in the 3-day embryo whereas three others are detectable only after hatching and may then undergo change until they reach the degree of agglutinability found in the adult. In *Columba guinea,* Miller (1953) found the antigens A,B,C,E, and F in 29-somite embryos at 68-72 hours of incubation. However, embryonic cells seem to be slower to agglutinate. The A and F antigens appear to develop gradually after hatching, for the cells of squabs containing A and F tend to show weaker agglutination and titer than the adult cells. Burke et al. (1944) prepared antisera against adult chicken red blood cells and carried out complement fixation tests with blood cells from successive stages of embryonic development. Complement was first fixed by cells from the 96-hour embryo. Unfortunately, interpretation of the latter results is doubtful since the data and experimental conditions are presented in very brief form and it seems that individual blood groups were not taken into consideration. Indeed the comparison of adult and embryonic cellular antigens is beset with technical difficulties because of the high degree of antigenic individuality of each bird and the fact that samples of blood are not taken from the same individual embryos. It has been possible to distinguish each chick in three different families of birds by means of the absorptive qualities of their red cells. Various aspects of blood cell serology in genetics and embryology are available in reviews by Irwin (1949, 1951).

Since the present evidence shows that at least some of the red cell antigens are epigenetic, we have no good reason for considering them any more or less significant as possible factors in differentiation than the

[6]

antigens arising in other tissues or at certain stages in the development of the sea urchin, frog, and chick. Moreover the blood cell antigens show the same general features as other tissue antigens. Some (like Rh) are specific to the red cells, whereas others (like A and B of man and K_5 of rabbits) are held in common with other tissues. Of course on the mere grounds of epigenicity and distribution we have no reason for attributing to them any causal role in the differentiation of the cells, but this applies with equal force to other antigens shown to arise during differentiation. In the case of the red cells, however, we are aware of a causal factor in the differentiation of the antigens. *The relationships of the red cell antigens to genic factors suggest that they are end products of differentiation rather than parts of the mechanism of differentiation.* We have at present no reason to assume that any other antigen of development is any different.

IV. EPIGENESIS OF HEMOGLOBINS

Chemical epigenesis has no better illustration than the hemoglobins. The first hemoglobin to appear is certainly a product of the embryo's own synthesis (autosynthetic), since unincubated eggs contain no iron heme (Hill, 1931). Ramsay (1951) states that nonheme iron accounts for 99 to 100.5 per cent of total iron in unincubated eggs. After the formation of this first hemoglobin there is a gap in our knowledge, for most of the work on fetal hemoglobins of vertebrates begins at the midfetal period. Within this advanced period in man, sheep, goat, cattle, and rabbit there is a large body of evidence for the existence of a fetal type of hemoglobin which is lost sometime after birth, and an adult type which gradually replaces it. The fetal type, at least in man, can be distinguished serologically from the adult type (Darrow, Novakovsky, and Austin, 1940; Chernoff, 1952). In addition the two hemoglobins can be distinguished on the basis of their solubility-temperature curves, their different crystal systems, electrophoretic mobility, amino acid composition, oxygen dissociation, resistance to alkaline denaturation, and other properties (Roughton and Kendrew, 1949, and Kendrew, 1949). The existence of an embryonic hemoglobin has not been shown in the chicken, but the studies of Hall (1934-1935) and Boyer (1950) on O_2 affinity and O_2 consumption of embryonic red cells are suggestive. Both sets of data are consistent with the assumption that an embryonic type of hemoglobin is gradually replaced by an adult type, although the evidence is insufficient to establish the fact for the chick embryo.

In the instance of the hemoglobins we have a tissue-specific and embryo-specific antigen (fetal hemoglobin) and the appearance of new antigens (fetal and adult-type hemoglobins) during the course of development. Their epigenetic and autosynthetic character is clear. It happens that we know something of the nature and functions of these antigens and their control by genic factors. Again the antigens are end products of differentiation.

V. EPIGENESIS OF THE BLOOD PLASMA PROTEINS

The plasma proteins of the developing organism show the epigenetic features apparent in blood cell antigens, hemoglobins, tissue antigens, and in histological and morphological differentiation generally. The evidence at the present time is sufficient to show that new proteins appear while others disappear, that some proteins are specific to the embryonic stages whereas others are continuously present throughout the life history. The relative proportions of the various components of the plasma are as characteristic as the changing proportions of the morphological units of the body.

The remarkable developmental changes in the plasma have not been apparent because so much of the work has been limited to the advanced fetus and postnatal life. In this period the general tendency is for the total protein to increase in concentration while the proportion of albumin decreases. Summaries of these investigations are given in Windle (1940), Barcroft (1947), and Smith (1951). In some instances the serum differences are very striking as in the newborn horse, which has little or no detectable γ-globulin before ingestion of colostrum, and the newborn rabbit, with only 0.2 per cent globulin (Haurowitz, 1950, p. 155).

However, these changes, important as they are, are largely of a quantitative nature without apparent *de novo* accessions or losses. The same may be said of most studies of the blood-clotting mechanism. It has long been known that the mammalian fetus has relatively long blood-clotting times and that the human newborn frequently requires 10-12 minutes. These differences have been correlated with quantitative changes in fibrinogen, calcium, prothrombin, or other elements of the clotting mechanism. (See Needham, 1931, and Smith, 1951, for reviews.) However, Pickering and Gladstone (1925) showed that the absence of clotting in chick embryo blood prior to the twelfth day of incubation is associated with the absence of both fibrinogen and prothrombin. At later stages the blood clots slowly because of quantitative differences in the clotting components. It seems clear that the proteins of the clotting mechanism show epigenetic development, but we must restrain any assumptions that the early clotting proteins are necessarily autosynthetic products.

Sherman (1919) could demonstrate complement in the chick's blood at 17 days of incubation but not in the unincubated egg. This is certainly an epigenetic component but it is not at all clear that it is synthesized by the embryo. In some of the earliest work Sachs (1903) found that several fetal pigs had only about one-twentieth of the complement present in adult pigs. On the other hand the newborn guinea pig has only slightly less complement than the adult. The events in man are in line with those in the guinea pig, as might be expected from the reduced type of placenta found in both. The human 14-week fetus shows low complement titers but the 28-week fetus shows titers close to those of the newborn and adult (Sölling, 1937). These data suggest that in the mammals complement, like antibody, is transmitted from the mother to the offspring according to placental type. The situation in the chick is not clear. From our knowledge of antibody transmission in the chick it would seem probable that complement also is a product of maternal synthesis. Sherman's failure to demonstrate complement in yolk requires that the subject be reinvestigated with emphasis on anticomplementary substances and "masked" complement of yolk and embryonic tissue.

The "natural" lysin for rabbit erythrocytes which occurs in chicken blood was found by Sherman (1919) in the 21-day chick (at hatching) but not in the 19-day chick. Rywosch (1907) also found that these lysins appear late in development, but shortly after hatching, according to her results. Again we have a clearly epigenetic event but the conclusion that the lysin is "natural" in the sense of human isoagglutinins is not warranted. As will be indicated below, immune factors transferred from the hen may be conserved in yolk and tissue of the embryo and released to the blood only in advanced stages of development.

Duran-Reynals (1940) showed that fowl serum contains substances, probably globulins, which flocculate saline tissue extracts of many species. These substances first appear 3 to 4 weeks after hatching when bacterial agglutinins are also appearing. They are of low specificity since they not only react with tissue components of many species but also may be increased by injecting a variety of antigens. The importance of this instance is that it illustrates the uncertainty involved in classifying certain serum constituents as "natural antibodies," presumably produced without the stimulus of homologous antigen. The factors described by Duran-Reynals seem to be antibodies of low specificity produced by active immunization of the chick.

In the case of the human isoagglutinins we are apparently on firmer ground when we think of them as natural components of the serum not requiring the stimulus of an exogenous antigen. Only about 50 per cent of newborn infants show such agglutinins although the human blood cell isoagglutinogens are present very early in development (Wiener, 1943). Isoagglutinin titers of the newborn are probably of maternal origin and, like many antibodies, enter the fetal circulation via the placenta; this is suggested by the fact that the congenital isoagglutinins disappear from the blood during the first two weeks after birth and their place is taken by the later isoagglutinins synthesized by the infant's body (see Thomsen and Kettel, 1929; Wiener, 1943).

It is evident from consideration of the clotting proteins, complement, "natural" lysins, and isoagglutinins that the mechanisms for their synthesis come into active production at relatively advanced stages of life. These macromolecules are certainly of epigenetic character but it is still doubtful that the first molecules which appear in the blood are products of the embryo's own synthesis despite their late appearance. The organism's own synthesizing mechanism in some instances come into play still later, and the heterosynthetic molecules are then replaced by those of auto-synthetic origin.

The serum proteins of the chick embryo have for obvious reasons been studied over a wider range of life history than in any other vertebrate. Serologically one can detect constituents with the antigenicity of adult serum fractions at successive stages of development. Thus the specific antigenicity of albumin, α-β-globulin, and γ-globulin are detectable in about 4 to 5, 6, and 9 to 12 days respectively (Nace, 1953). The earliest blood samples obtained in sufficient quantity for serological analysis (3 to 4 days incubation) react with the antisera versus adult serum. If the antiadult serum is absorbed with yolk its reactivity for these early blood samples disappears completely. Nevertheless the antisera still react with blood extracts from the fifth day of development and with all older embryos. This has been interpreted to mean that the very early blood is composed largely of one or more of the soluble proteins present in that fraction of the yolk known as livitin (Nace and Schechtman, 1948). Serum proteins having the antigenicity of adult proteins but not of yolk proteins (nonvitelloid constituents) are regularly detectable on the fifth to sixth days of incubation. It is of course possible that they occur in the earlier blood but in concentrations too low to react with the antisera. Nevertheless it is remarkable that the blood shows this antigenic change at a time when the primitive blood cells are disappearing and large numbers of disintegrating cells are seen in the circulating blood. At this same stage substances having the specific antigenicity of the α-β-globulins of adult blood are also detectable for the first time (Schechtman and

Hoffman, 1952). Components with the electrophoretic mobility of a-β-globulins are prominent in extracts of various adult tissues; their rise in the embryonic blood coincident with extensive cytolysis supports the view that disintegrating blood cells make an appreciable contribution to the plasma proteins. This had been suggested earlier from observations on plasma formation in the blood islands of the chick (Sabin, 1920) and a good deal of evidence indicates that similar processes occur in some adult mammals (Sabin, 1939; White and Dougherty, 1946). It should be noted that none of these considerations establishes any presumption in favor of the autosynthetic character of the substances released.

Proteins antigenically similar to or identical with certain yolk proteins occur in chicken blood throughout the entire life history. Antisera prepared against the water-soluble (livetin) fraction of yolk show clear reactions with the serum of 80-hour embryos (the earliest stages from which blood uncontaminated by other tissues could be obtained) and with all other developmental and post-hatching stages tested, including adult males and females (Schechtman, Nace, and Nishihara, 1953). The same conclusions are indicated by the use of antisera against adult serum fractions (Nace, 1953). Marshall and Deutsch (1951), using quantitative precipitin tests, have presented strong evidence that ovalbumin of egg white is very similar to yolk albumin and to adult serum albumin. Also conalbumin of egg white is very similar to and perhaps identical with components found in both yolk and adult serum. Finally proteins similar to or identical with conalbumin and ovalbumin are present in the serum of the 13-day chick embryo on the basis of both electrophoretic and serological studies. Ovomucoid and lysozyme of egg white could not be detected in the yolk extract, as prepared by Marshall and Deutsch, nor in embryonic serum or adult serum.

The most abundant type of protein in the yolk is a phosphoprotein (or complex of several phosphoproteins) designated as vitelline, ovovitelline, or lipovitelline. This is probably completely hydrolyzed by the embryo, since little if any *protein* phosphorus can be found in the embryonic serum (Marshall and Deutsch, 1950). It is either absent or present in very low concentrations in the immature hen and the cock but is easily identified in the serum of the laying hen (Roepke and Bushnell, 1936). The vitelline content of adult blood can be increased in immature birds by injections of androgens or estrogens (Common, Rutledge, and Bolton, 1947). We have recently repeated experiments like those of Roepke and Bushnell and found that it is possible to identify the serum of the laying hen by its serological reaction with antivitelline

(Schechtman, Nace, and Nishihara, 1953). In addition antivitelline sera were found to react with embryonic serum from the earliest samples tested (80-hour chick) throughout embryonic development and post-hatching life. However, since it was impossible to obtain an antivitelline which would not react with livetin, even after extensive absorption, it must still be considered most probable that phosphoprotein is probably not present in the serum in appreciable concentrations until sometime before egg laying commences.

Among the electrophoretic components of embryonic chick serum there are two or three which migrate ahead of albumin and which vanish by the third day after hatching (Marshall and Deutsch, 1950). These fast proteins have a high phospholipid content. Their loss shortly after hatching would indicate they are embryonic proteins in the same sense that certain structures like the notochord of higher vertebrates are embryonic. For a while their embryonic character was doubtful since Brandt, Clegg, and Andrews (1950) found a faster-than-albumin component in the serum of the laying hen by the use of borate buffer rather than the phosphate and veronal buffers more commonly used. They could not find the fast component in sera of the cockerell and nonlaying hen. Heim (1953) has used the borate buffer and has verified the findings of Brandt, Clegg, and Andrews on adult birds. The same buffer was used in electrophoretic studies of the serum proteins from the eighth day of incubation to the adult. It was found that two or three faster-than-albumin components are present throughout the embryonic period. This does not of course establish the identity of the fast embryonic and adult components. Heim has pointed out that the components can hardly be identical since veronal buffer brings them out in the embryonic but usually not the adult serum, whereas borate buffer is effective for both. Moreover the faster-than-albumin components have thus far not been discernible in the electrophoretic patterns of yolk extracts and egg white.

The evidence of embryo-specific proteins is not limited to electrophoretic studies. Serological tests indicate the same conclusion. Rabbit antisera against the serum of 10-day embryos were absorbed to completion with adult laying-hen serum. Such antisera retained activity for the 10-day embryo serum (Schjeide, 1953). Specific anti-10-day sera reacted with all embryonic sera from the tenth incubation day to several days after hatching. The correspondence of these serological results with the electrophoretic data of Marshall and Deutsch, and Heim, is most striking.

[12]

Quantitative precipitin tests with antisera versus 10-day embryo chick serum show that some of the antigenic constituents of the serum at this stage are radically different from those of the adult (Piccarillo, 1952). One has no difficulty in distinguishing the precipitin curves of the two. The same applies to antiadult sera reacted with embryonic and adult sera. These curves tell us nothing as to the nature of the embryonic components; they simply indicate antigenic differences. It would seem probable, however, that much of the difference between the embryonic and adult sera is attributable to the faster-than-albumin, the α-β-components, or both; since Marshall and Deutsch (1951) find such a close similarity between the albumin and the γ-globulin regions of the electrophoretic pattern.

Doubtlessly the best evidence for an embryonic protein is found in the work of Pedersen (1944, 1947) on the physical and chemical characteristics of fetuin, a protein of low molecular weight (about 50,000) with an electrophoretic mobility corresponding roughly to that of α-globulin. In the advanced beef fetus it makes up 34 per cent of the total protein; in the 5 to 6 month calf it has dropped to about 5 per cent, and in the adult there are no more than traces. Although albumin has a low isoelectric point in comparison with common globulins, that of fetuin is still lower (3.5 for calf fetuin). Since its sedimentation constant is low (e.g. cow γ-globulin, 7.5; human albumin, 4.6; calf fetuin, 3.28 Svedbergs) and it contains an appreciable amount of P, it may be inferred that it is a lipoprotein. Both the sheep and beef fetuses contain fetuin but human cord blood and fetal rabbit blood do not; hence this protein is possibly of physiological importance only in mammals with primitive (unreduced) types of placentae.

Insofar, then, as the more abundant serum proteins are concerned (e.g. those components detectable by electrophoresis), it is evident that the serum shows complex epigenetic development. The embryonic constituents such as the faster-than-albumin components of the chick and fetuin of cattle and sheep would seem of necessity to be autosynthetic. The albumin and globulins may be heterosynthetic, autosynthetic, or of double origin at various stages of development. The close similarity or even identity of the albumin and γ-globulin components of the adult blood, yolk, egg white, and embryonic serum does not of course establish the adult alone as the site of their synthesis. Nor does it indicate that these proteins are synthesized first in the adult, then within the developing ovum, and again in the embryonic chick. It is only in the case

[13]

of macromolecules marked by their specific combining properties (anti-bodies) that we have good evidence for the adult origin of some of the embryo's proteins.

VI. LATE DEVELOPMENT OF THE MECHANISM OF ANTIBODY SYNTHESIS

It is well known that the embryonic stages of life are generally char-acterized by deficiency in the mechanism of antibody production. This is reviewed by Needham (1931) and Wolfe and Dilks (1948) and has been corroborated many times in recent years. Thus Wolfe and Dilks (1948) found that about 50 per cent of newly hatched chicks show weak antibody production after injections of bovine serum, and the remainder show no detectable antibody production. Antibody titers increase gradu-ally up to the fourth week of posthatching life; then in the 5-week-old chick there is a great increase over the 4-week specimens, and the mecha-nism of antibody production is essentially mature. Brandly, Moses, and Jungherr (1946) inoculated Newcastle virus into the yolk and allantois of chicks from the sixth to the eighteenth days of incubation. There was no evidence of active antibody formation by the embryo. Although the eggs were derived from hens immunized against the virus and such eggs contain high antibody titers in the yolk and embryonic tissue, the chick *blood* lacked antiviral immunity until about the fifteenth day of incuba-tion after which it increased progressively. This is when passively ac-quired immunity, transferred from the hen, appears in the embryonic serum. In recent extensive studies of a cattle virus (Baker and Greig, 1946; Shope and Griffiths, 1946) it was shown that the 7-day chick embryo is an excellent medium for culture of the virus. Injected into the yolk sac, the virus can be harvested 4 days later from both the embryo and the extra-embryonic fluids. If the chicks are allowed to develop the virus can be found in the brain and solid viscera 5 days after hatching. Yet antiviral activity is not detectable until 3 or 4 weeks after hatching. If 1-day chicks (posthatching) are infected there is no antiviral activity at 14 days, but at 21 days about 85 per cent of the sera are antiviral, and by the twenty-eighth day the sera can neutralize ten times as much virus as the 21-day sera. In brief, irrespective of when it receives the virus the chick attains the capacity to produce antiviral substances 3 to 4 weeks after hatching. The use of the chick embryo for the cultivation of various bacteria, viruses, and heterologous normal and neoplastic tissues presumably depends to a large degree upon its inability to synthesize the highly specific immune proteins (antibody).

Many and perhaps all of the immunities attributed to the young chick and to very young animals in general doubtlessly do not depend upon the synthetic mechanisms of the young organism itself but upon presynthesized molecules transmitted from the mother. As we have seen, one may reasonably be skeptical even of the "natural" lysins and of complement which appear late in chick development as well as the many bacterial immunities known in the young of domestic animals. The idea that high metabolic rate and rapid growth in some way create a general immunity is no doubt an illusion which arose from observations of congenital immunities without considering the immunities present in the mother's blood and colostrum. The economic importance of certain immunities in domestic animals during the first few weeks of postnatal life has already given us a good deal of evidence on the deficiency of the antibody-forming mechanism in the embryo and young organism. This work has been reviewed by Brandly, Moses, and Jungherr (1946) and more recently by Bornstein et al. (1952). Antibody formation is one of the last major physiological processes developed.

VII. ANTIGEN UPTAKE BY THE EMBRYO

The absence of antibody formation in the embryo can hardly be based upon any failure of the embryonic cells to phagocytize or arthrocytize antigens. Heine (1936) found that all three germ layers of the early chick embryo, including the red blood cells, could phagocytize India ink. In later stages phagocytic activity is progressively delimited to the familiar sites of the mature organism. Since Heine's injections caused stoppage of the circulation and may therefore have induced trauma, Steinmuller (1937) repeated and extended the work. India ink injected under the blastoderm was taken up by the endoderm; but since it was also found in the ectoderm it must have been transferred from cell to cell. In 2-to-4-day embryos both the thin centrally-located endoderm and thick syncytial endoderm were phagocytic, and India ink was visible in the cell bodies and protoplasmic extensions of the mesenchyme.

The chick embryo also shows good phagocytosis in advanced stages of development, but it is now restricted to specific types of cells. Goodpasture and Anderson (1937) inoculated various pathogenic bacteria into the chorioallantoic membrane and observed extensive phagocytosis. They gained the impression that survival of certain pathogens is favored by the intracellular environment. In the 12-day embryo Buddingh and Polk (1939a,b,c) found that the endothelial cells of the chorioallantois have a specific affinity for meningococci. In 15-to-16-day embryos many large mononuclears come out of the connective tissue and phagocytize the cocci. After injection into the blood stream of 15-day embryos the phagocytized cocci seem to multiply within the phagocytes. Canat and Opie (1943) showed that 3-day chick embryos ingest various types of particulate matter including erythrocytes. It was not until the very advanced stages of development that granulocytes took any important part in the inflammatory response. Buddingh and Polk (1939a,b,c) observed that in the advanced embryo, cells of mononuclear type are more active in phagocytosis than are granulocytes. Therefore it is obvious that at least some antigens can enter embryonic cells. In the case of meningococci, certain staphylococci, and streptococci, the bacteria may even be destroyed, apparently by humoral factors. It is of interest that signs of antibacterial activity appear on the fourteenth to fifteenth day, which is the period when immunities known to be transmitted from the hen first become detectable in the blood. Nevertheless, active antibody formation (autosynthesis) does not occur. This suggests that the embryo actually lacks the specific mechanism of antibody synthesis.

VIII. TRANSFER OF SERUM PROTEINS IN BIRDS AND AMPHIBIANS

Shortly after Ehrlich (1892) showed that immunity in the mouse is transmitted to the fetus *in utero* and by way of mammary secretions after birth, Klemperer (1893) showed that the yolk but not the white of the hen's egg has antitoxic properties if the hen has been immunized previously with the specific antigen. Since then the deposition in the yolk of protective factors against toxins and various infectious agents has been amply verified. Meanwhile it was shown that the specific antibody activity of the yolk later appears in the blood and in some tissues of the embryo, and may persist for a number of weeks after hatching. The third main step was the localization of antibody activity in the water soluble proteins of the yolk (Dzierzgowski, 1901) and their association with the hen's serum globulin and a corresponding protein fraction of the yolk (Jukes, Frazer, and Orr, 1934; Frazer et al., 1934). This literature is reviewed in Brandly, Moses, and Jungherr (1946), Sherman (1919), Romanoff (1949), and Bornstein et al. (1952).

It is important to note that antibody is transferred to the chick in a manner which insures protection during the first weeks after hatching with little expenditure in the first three-fourths of embryonic life. Whether this is generally true remains to be seen, but in eggs from birds immunized against Newcastle disease virus appreciable antibody is not demonstrable in the serum until sometime between the twelfth and fifteenth days of incubation, although the yolk and embryonic tissue show high titers (Brandley, Moses, and Jungherr, 1946). Both the antiviral activity of the embryonic serum and the resistance of the chicks increase during the last quarter of the developmental period. Bornstein et al. (1952) using hemagglutination-inhibition tests showed that the titer at 18 days of incubation is low compared with that at 2 to 3 days after hatching. Now if we knew nothing about the transfer of the antibody from the mother and its presence in the yolk and tissues it would be quite logical to conclude that some chick embryos begin to synthesize a natural antibody in late embryonic life.

In amphibians the transmission of antibody or serum proteins via the yolk has apparently not been shown or perhaps not even attempted. Studies of this kind will encounter difficulties since amphibia do not readily form antibodies and their yolk is intracellular. The demonstration by Bisset (1947) that frogs maintained at 20°C. will develop agglutinating antisera against *Pseudomonas fluorescens* suggests that the serological approach may possibly be put to use in the amphibia. However, the intracellular yolk of amphibians presents a serious problem. We have a concrete entity in mind when we speak of bird yolk, but the situation is otherwise when we turn to the amphibian. Intracellular yolk is a rather nebulous concept and the definite material which various authors have segregated by centrifugation is uniform in name rather than substance. Where shall we draw the boundary between yolk granules and cytoplasmic particulates in amphibian eggs? Shall we centrifuge briefly and obtain only the coarsest granules, or for a longer time and include mitochondria and microsomes? Is it still yolk after we have washed it well in saline solution and thereby removed a disproportionate fraction of the saline soluble proteins? Holtfreter (1946) describes two main types of yolk inclusions, the small lipochondria and the large vitelline platelets, both broken up by the same reagents in spite of some differences in chemical constitution. Washing yolk is therefore likely to change its constitution radically, for the substance of the small lipochondria will tend to disappear in the wash and the proportion of coarser particulates will increase. Furthermore if we wish to make comparisons with avian yolk we are faced with the highly probable fact that centrifugal force applied to the amphibian embryo is unlikely to throw down the same materials which the bird oocyte naturally segregates. For example in preparing amphibian yolk I see little reason for excluding the floating fat-rich layer of relatively low specific gravity; such a procedure applied to bird yolk would certainly result in loss of important lipoprotein fractions.

[17]

The oocytes of *Rana temporaria* contain antigens like those in the serum of the adult as shown by Flickinger and Nace (1952). Later stages (the blastula, neurula, and tailbud) show similar antigens. Cooper (1948) obtained antisera against the washed yolk-platelet fraction of frog eggs and showed that it reacted with adult serum. Considering the various extracts and antisera used by Cooper and by Flickinger and Nace there can be little doubt that antigens very like the serum proteins are present throughout the life history of the amphibian. Whether these antigens simply contain some antigenic determinants in common with serum proteins, or are actually serum proteins, remains to be seen. The application of the quantitative precipitin test or further studies of the physical and chemical properties of amphibian yolk proteins may provide the necessary evidence.

IX..TRANSFER OF SERUM PROTEINS IN THE MAMMAL

The transmission of presynthesized macromolecular constituents to the mammalian embryo via the yolk sac in stages equivalent to the primitive streak chick embryo and amphibian gastrula has been demonstrated only recently. This was preceded by studies which indicated a high degree of permeability in the blastocyst. Everett (1935), on the basis of his studies on permeability of the rat blastocyst and later stages, pointed out that the role of the yolk sac has been underestimated. Certain dyes perfused into the uterine vessels of the pregnant rat readily penetrate the blastocyst wall and can be seen in the cavity and epithelium of the yolk sac. Thus trypan blue was found within Reichert's membrane and in the cells of the yolk sac epithelium but iron ammonium citrate was excluded from the yolk sac. Histochemical studies such as those of Wimsatt (1949) have provided strongly suggestive evidence that the yolk sac endoderm of at least some mammals has absorptive functions but as pointed out by Wimsatt the point has not been established on the basis of histochemical studies. Unfortunately the yolk sacs in histological sections frequently give the impression of being empty. No doubt the soluble constituents are in part dissolved out or form a negligible precipitate.

It was largely by means of serological methods that Brambell and his colleagues were able to obtain convincing evidence for the transfer of maternal serum proteins into the rabbit blastocyst at the time when the basic plan of the body is being established. The experiments and much related work have been summarized by Brambell, Hemmings, and Henderson (1951). Agglutinins for *Br. abortus, whether actively produced by the rabbit or passively obtained by injections of bovine antiserum,* pass into the yolk sac cavity of blastocysts on the seventh and eighth days *post coitum*. At this time the endoderm is spreading over the inner surface of the trophoblast and the mesoderm is extending outward from the primitive streak. On the eighth day the amniotic folds are forming and beginning to invade the uterine mucosa and blood vessels are developing in the splanchnic mesoderm of the yolk sac. During this period the volume of the yolk sac cavity increases about seventy times.

By the ninth day the blastocyst has changed its permeability and now the antibodies of the maternal blood enter the yolk sac only to a slight degree or not at all. It is interesting to note that the blood vessels of the yolk sac are developing on the eighth day when the blastocyst is still permeable to heterosynthetic protein. By the ninth day the heart begins to beat, the vitelline circulation is established, and now antibody protein no longer enters the yolk sac.

Insofar as the relationship between growth of the blastocyst cavity and the development of blood vessels is concerned, the same relationships are common and perhaps universal in mammals. Thus in the human blastocyst the yolk sac reaches its maximal size (2 mm. diam.) at about the start of the fourth week, when the blood vessels and blood islands are rapidly differentiating in the wall of the yolk sac. By the end of the fourth week the heart beat starts and circulation is established.

The permeability of the early blastocyst should not be confused with the permeability of the placenta or yolk sac remnants in advanced development. The classification of placentae according to the number of layers separating the maternal from the fetal circulation (Grosser, 1909, 1927) on the whole coincides well with their permeability to antibodies, multilayered placentae being associated with relative impermeability (Kuttner and Ratner, 1923; Mossman, 1926). A number of considerations which may eventually require modification of this concept are discussed by Brambell, Hemmings, and Henderson (1951). In any event the Grosser classification has no pertinence whatsoever to macromolecular transmission in the young blastocyst. The placental types have not yet been elaborated and the transmission that does occur is between the maternal blood (or tissue?) and the fluid contents of the blastocyst. Whereas the advanced placentae or fetal membranes of mammals show marked differences as to antibody transfer, we have no evidence that such differences exist in the early blastocyst stage when the embryo is gastrulating. On the contrary the small available evidence favors the view that the early blastocyst is probably highly permeable in all mammals. For example in the rat the transfer of antibody is largely via the colostrum and milk

(Culbertson, 1938; Kolodny, 1939; Fenner, 1948) although some passes *in utero*. On the other hand in the rabbit essentially all the antibody is transferred before birth and little or none after birth. Yet the blastocysts of both species show a high degree of permeability insofar as available evidence is concerned. Thus trypan blue (Everett, 1935) and ferric ammonium citrate (Brunschwig, 1927) pass into the rat yolk sac despite the fact that it is lined with Reichert's membrane, which is absent in the rabbit.

Another point which illustrates that one can not make assumptions about the permeability of the blastocysts from that of the advanced placenta is to be found in the work of the Brambell school on the rabbit itself. In advanced development, at about 20 days, both rabbit-made and bovine-made antibodies are admitted into the fetal circulation. From this time to the twenty-fourth day a selective mechanism enters the picture: rabbit antibody is now taken up as readily or more readily than at the twentieth day, but bovine antibody is almost entirely excluded.

Most work on macromolecular transfers has been focused on the passage of antibodies simply because these are easily identified by their specific avidity for the antigen. At the same time the general significance of the transfers has been given little attention probably because the functions of antibodies seem so obvious. However, Brambell et al. (1949) showed by electrophoretic analysis that the contents of the early yolk sac of the rabbit contains components with mobilities corresponding to all the main components found in the maternal blood plasma (albumin, γ- and β-globulins, γ-globulin, and fibrinogen). In the ultracentrifuge the yolk sac fluid showed two components which sediment, as do those of the adult plasma, and no materials of low molecular weight. The blastocyst wall seems to exercise little discrimination as to the type of protein that enters. This is shown by the fact that the percentage composition of the yolk sac constituents is essentially the same as that of the maternal plasma although total concentration is lower. The *transfer of macro-molecules is thus not limited to antibody globulin and we need no longer suppose that transmitted macromolecules are involved only in establishment of the immune state.* We can not at present ascribe any other definite function to these transmitted proteins other than that suggested by the serological studies of the chick embryo's blood, namely that one or more of them is translocated at various times in development to the circulation where they become constituents of the plasma. Nace (1953) has suggested that they may serve as templates for the elaboration within the embryo of corresponding serum proteins. As will be indicated below the behavior of the early mammalian embryo *in vitro* and in transplantation suggests the further possibility that transmitted macromolecules may well have a differentiative function. Of course it is possible that the mammalian serum components remain in the yolk sac, but I think this rather improbable from the known fate of the reptilian and avian yolk sac contents and the readiness with which the mammalian yolk sac entoderm takes up colloidal matter.

From the electrophoretic analyses of the fluid in the rabbit yolk sac (Brambell et al., 1949) the mammalian yolk sac proteins are not equivalent to the avian yolk but to its livetin fraction. The electrophoretic analyses of chicken livetin (Shepard and Hottle, 1949) show that it contains components corresponding in mobility to albumin, β-globulin, and γ-globulin of the adult serum. Therefore we should view with considerable caution the concept that the mammalian yolk sac is a nonfunctional vestige of reptilian ancestry.

Certainly the embryonic serum is not merely a dilute adult serum of lower protein concentration. Embryonic serum at all stages of development studied has distinct physical and chemical properties. Some of these have been reviewed recently (Schechtman, 1952). As to the most abundant proteins, electrophoresis shows clearly that they do not have the percentage composition of the adult serum. For one thing there is evidently a selective process by which some proteins are thrown into the circulation whereas others are retained longer in the yolk or tissues of the chick, e.g. the late arrival of anti-Newcastle virus activity in the peripheral blood. Similar selective processes are operative in advanced stages of the rabbit. In addition the embryo's own synthesizing mechanisms gradually mature and contribute to the plasma.

X. HETEROSYNTHESIS AND EARLY DIFFERENTIATION

We have seen that the ovarian egg of the hen, the early blastocyst of the rabbit, and the advanced embryos of birds and mammals show a remarkable degree of permeability to macromolecules. Particularly for the oocytes the cytological literature contains descriptions which strongly suggest the transfer of particles from somatic tissue to the germ cells. Painter (1940) described the uptake of chromosomes or entire nurse cells by the oocytes of *Drosophila* and suggested that this may be the basis of matroclinous inheritance. Schultz (1952) has discussed the potential significance of incorporation of nuclear material into the cytoplasm of the oocyte. He thinks it may provide a concrete mechanism for the transfer of "partial gene replicas." Brambell (1926) in a cytological

[21]

study of oogenesis in a fowl described complex particles which seem to pass from nurse cells to oocytes.

Now if the period of yolk accumulation (oocyte growth in amphibians and birds, blastocyst growth of mammals) is characterized by permeability to antibodies and perhaps particulates, we may reasonably expect other heterosynthesized macromolecules to enter the young organism. This assumption is supported by the work of McCance et al. (1949), who found that during the period of antibody absorption in the dog the gut is also highly permeable to at least one enzyme. Antibody transfer in the dog occurs mainly via the colostrum, which may exceed the maternal serum in antibody titers. Now dog colostrum also has an unusually high cholinesterase activity and this is transferred to the blood of the suckling young. Control puppies fed evaporated cow's milk did not show the high cholinesterase activity. This finding presents new prospects to embryological study, for if antibody permeability is correlated with enzyme permeability during early stages—a point of great importance yet to be established—there is the possibility that the mother supplies the oocyte or blastocyst with enzymes, nucleic acids, and a variety of macromolecules other than the common serum proteins. It suggests the possibility that the large amount of desoxyribosides found in the frog egg eytoplasm (Hoff-Jorgensen and Zeuthen, 1952; Zeuthen, 1951) may be taken up in macromoecular form and have functions other than that of supplying building blocks for the nuclei of the embryo.

The passage of maternal serum into the blastocyst where it forms the immediate environment of the mammalian embryo provides us with a clue to the problem of the differentiation requirements of the early mammalian embryo. Although we have a good deal of evidence concerning the factors necessary for tissue proliferation *in vitro,* we know next to nothing concerning the conditions which "permit" the mammalian gastrula to enter the period of differentiation. What is worse, it is not generally realized that *in vitro* conditions which are perfectly suitable for the proliferation of tissues, and even for the differentiation of rudiments from the advanced embryo, are not suitable for the basic differentiation which follows the gastrula stage. In stating that certain conditions "permit" this basic differentiation, we are expressing a current and unfounded assumption as to the nature of the mechanism of differentiation: the assumption that the medium surrounding the embryo simply provides the "conditions of life" and that the mechanisms of synthesis and differentiation are elaborated within the nuclei and cytoplasm of the embryo. This assumption has doubtlessly arisen from the great

[22]

amount of experimental work carried out on the amphibian and chick embryos which apparently obtain a supply of serum constituents from the mother in the oocyte stage and are thus provided with presynthesized macromolecules during the critical early stages of differentiation. Hence we find little difficulty in securing the basic differentiations when using early stages of the amphibians and birds. Rudnick (1938) explanted portions of the chick blastoderm at 16 to 22 hours of incubation onto the surface of a clot composed of plasma and embryo extract and obtained neural plates, notochord, body wall, and coelomic vesicles. Spratt (1948) showed that the chick primitive streak stage will carry out its basic differentiations in a medium containing no organic material other than a simple sugar (e.g., glucose). As was previously pointed out (Taylor and Schechtman, 1948) this does not indicate that the work of differentiation was accomplished solely on a sugar substrate since the cells of the early blastoderm are rich in intracellular yolk.

In the mammals the transfer of maternal macromolecules takes place at a relatively advanced period, during the growth of the blastocyst. It is not surprising therefore that removal of the mammalian embryo prior to or during this critical period of macromolecular transfer creates deficiencies not encountered in the amphibians and birds.

Students of early mammalian differentiation have been only too well aware of these differences. Nicholas (1937, 1947) has pointed out that *in vitro* conditions suitable for proliferation are not necessarily also suitable for early differentiation and that the processes of differentiation and growth can be separated by explantation of the mammalian embryo into tissue culture. Nicholas (1937) has also pointed out that the fish and amphibians can develop in inorganic media since they rely upon cell inclusions which are deficient in the mammal. By providing the 9-day rat embryo with a circulating medium rather than the still media more commonly used, Nicholas obtained improved differentiation. It is of interest that homologous plasma was not necessary, cat plasma serving as well as rat plasma. This is in line with the work of Brambell and his colleagues, who showed that bovine antibodies as well as rabbit antibodies readily enter the rabbit blastocyst and the circulation of the advanced fetus at 20 days *post coitum*. However Nicholas could not consistently repeat the conditions necessary for differentiation since the nature of the critical factors is insufficiently known.

Grobstein (1950–1952) obtained similar results with mouse embroyos explanted during the stage of mesoderm formation. The trophoblastic portion of the blastocyst was removed in dilute horse serum and the embryonic shield was placed into a medium which we know to be an excellent one insofar as proliferation is concerned (chicken plasma, horse serum, chick embryo extract, and Tyrode solution). However, the youngest embryonic shields disintegrated without either growth or differentiation. Slightly older shields fared better but still showed little capacity for differentiation. When the embryo has reached the head-fold stage (cylinders about 1.0 mm. long) it has attained a fair degree of stability, for it is now capable of some spreading and differentiation under the *in vitro* conditions employed. We see here a phenomenon familiar to the amphibian embryologist. The embryo is competent to respond to differentiation promoting factors during limited periods; once a given competent period has passed it can no longer respond. In the mammals, however, we are concerned with a new order of competence (general differentiation) which has presumably escaped notice in amphibians and birds because the "inductors" are always present in the latter from the oocyte stage.

There is no evidence that I am aware of to indicate that *early differentiation* requires materials specific to embryonic stages, although it is well known that *proliferation* is favored by such factors. Thus Grobstein maintained clusters of 4 to 9 embryonic shields in a single culture and found that the greater quantity of tissue is conducive to the maintenance and spread of the cultures but does not contribute to their capacity for differentiation. Nicholas (1937) found that the addition of adult cat or rat plasma to pig amniotic fluid gave best results in growth and differentiation; the amniotic fluid itself was without effect. Nicholas and Rudnick (1933) transplanted whole rat embryos or pieces upon the chick chorioallantois. The critical gastrulating stages are hardly capable of survival. 5-day, 7-day, and 8-day embryos did not survive and only 2.9 per cent of 8.5-day embryos showed differentiation. With increasing age of the embryos the proportion of differentiated specimens increases to 15 per cent at the 10-day stage. In these advanced stages the organ rudiments have been established. It was not the environment of chick embryonic tissue which promoted differentiation but processes which had transpired in the rat embryos prior to their removal from the mother.

The necessity of *adult* materials for early differentiation is shown by transplantation of early mouse and rat embryos into various positions in the adult body. We know from the experiments of Nicholas that the rat embryo can differentiate without its fetal membranes and in many positions other than the uterus. Thus the early rat embryo shows excellent differentiation under the kidney capsule of the adult rat (Nicholas, 1942). Transplanted 2 to 4 cell stages and egg cylinders carried out a variety of histological differentiations. It was noted that a serum-like fluid, sometimes containing blood, formed around the transplants and was followed on the next day by a "surge of differentiation" as well as the spreading of the embryonic tissue over the face of the kidney. Waterman (1936) found that embryonic rabbit tissue gives numerous differentiations after transplantation into the adult animal but behaves poorly on the chick chorioallantois, where differentiation is limited to the rudiments already established. Grobstein (1951) found that young mouse embryos find a favorable site for differentiation in the anterior chamber of the adult mouse eye. Even embryos in mesoderm formation (0.4 to 0.6 mm. cylinder length) show moderate growth and differentiation of nervous and hepatic tissues in 20 to 40 per cent of the cases. If the young embryonic shields are first maintained *in vitro* and then implanted into the eye, they show a very limited differentiation capacity. But if the embryos at the time of explantation have reached approximately the head process stage, they are inhibited to a much lesser degree by *in vitro* conditions.

It seems probable from the above experiments with gastrulating mammals that *serum and fluids of the adult body supply factors necessary for the basic differentiations which follow the gastrula stage. But it does not follow that the same presynthesized molecules are necessary after the establishment of the main organ rudiments.* The decreased permeability of the rabbit blastocyst on the ninth day, when the vitelline circulation begins, suggests that the embryo is partly or entirely closed to macromolecules at this time. Moreover the available evidence indicates the existence of a selective mechanism for the intake of macromolecules, which for example brings antibody into the chick's circulation at about the fifteenth day of incubation and closes the rabbit fetus to bovine antibody on the twenty-fourth day of development.

XI. SUMMARY

It has been pointed out that sufficient evidence is available at the present time from serological as well as physicochemical studies to support the conclusion that differentiation at the molecular level is of epigenetic character. Moreover in its broad aspects molecular differentiation shows the same epigenetic features as do histological and organological development. In no case has it as yet been demonstrated that the new antigens or other chemical entities which appear in the course of development are components of the mechanisms of differentiation; where sufficient evidence is available the new molecular differentiations appear to be equivalent in character to the better-known morphological events—end products of differentiation.

Largely on the basis of serological studies, it is pointed out that molecular differentiations differ from morphological differentiations in at least one main aspect: certain of the epigenetic antigens or other macromolecules are not the end products of the embryo's own synthesis but are presynthesized molecules elaborated by the maternal tissues and transmitted to the embryo. Some evidence suggests that the early blood plasma of the embryo and certain constituents of advanced embryonic blood are derived wholly or in part from such heterosynthesized maternal macromolecules. Comparison of the differentiative capacities of amphibian, bird, and mammalian gastrulae *in vitro* and in transplantation suggests that the basic differentiations of the postgastrular period depend upon the presence of heterosynthesized macromolecules. It remains for future research to show whether the macromolecules of adult origin serve merely as *end products* of differentiation or take an active part as inductors or transformation factors in the *mechanism* of differentiation.

BIBLIOGRAPHY

Andresen, P. H. 1947. Blood groups with characteristic phenotypical aspects. *Acta Path. Microbiol. Scand. 24,* 616-618.

Baker, J. A., and A. S. Greig. 1946. Rinderpest. XII. The successful use of young chicks to measure the concentration of Rinderpest virus propagated in eggs. *Am. J. Vet. Res. (part 2), 7,* 196-198.

Barcroft, J. 1947. *Researches on Prenatal Life.* Charles C. Thomas, Springfield, Ill.

Bisset, K. A. Effect of temperature on immunity in amphibia. *J. Path. Bact. 59,* 301-306.

Boivin, A., A. Delaunay, R. Vendrely, and Y. Lehoult. 1945. L'acide thymonucleique, polymérisé, principe parraissant susceptible de déterminé la specificité serologique et l'équipement enzymatique des bactéries. *Experientia 1,* 334-335.

Bornstein, S., and M. Israel. 1942. Agglutinins in fetal erythrocytes. *Proc. Soc. Exp. Biol. Med. 49,* 718.

Bornstein, S., A. Rautenstein-Arazi, and Y. Samberg. 1952. Some aspects of congenital passive immunity to Newcastle Disease in chicks. I. The transfer of hemagglutinin inhibitors from maternal yolk to the chick. *Am. J. Vet. Res. 13,* 373-378.

Boyer, C. C. 1950. Respiration of embryonic blood. *Proc. Soc. Exp. Biol. Med. 75,* 211-214.

Brachet, J. 1949. L'hypothèse des plasmagenes dans le developpement et la differentiation. *Pub. Staz. Zool. Napoli 21*, 77-105.

Brambell, F. W. R. 1926. The oogenesis of the fowl (*Gallus bankiva*). *Trans. Roy. Soc. Lond.* (B) *214*, 113-151.

Brambell, F. W. R., W. A. Hemmings, E. F. McCarthy, and R. A. Kekwick. 1949. Passage into embryonic yolk sac cavity of maternal proteins (plasma) in rabbits. *J. Physiol. 108*, 177-185.

Brambell, F. W. R., W. A. Hemmings, and M. Henderson. 1951. *Antibodies and Embryos*. University of London, the Athlone Press.

Brandly, C. A., H. E. Moses, and E. L. Jungherr. 1946. Transmission of antiviral activity via the egg and the role of congenital passive immunity to Newcastle disease in chickens. *Am. J. Vet. Res. 7*, 333-342.

Brandt, L. W., R. E. Clegg, and A. C. Andrews. 1950. Comparisons of the serum proteins of young and mature chickens. *J. Biol. Chem. 191*, 105-111.

Briles, W. E., W. H. McGibbon, and M. R. Irwin. 1948. Studies of time of development of cellular antigens in the chicken. *Genetics 33*, 97.

Brunschwig, A. E. 1927. Notes on experiments on placental permeability. *Anat. Rec. 34*, 237-244.

Buddingh, G. J., and A. Polk. 1939a. Experimental meningococcus infection of the chick embryo. *J. Exp. Med. 70*, 485-498.

Buddingh, G. J., and A. Polk. 1939b. The pathogenesis of meningococcus meningitis in the chick embryo. *J. Exp. Med. 70*, 499-510.

Buddingh, G. J., and A. Polk. 1939c. A study of passive immunity to meningococcus infection in chick embryos. *J. Exp. Med. 70*, 511-520.

Burke, V., N. P. Sullivan, H. Petersen, and R. Weed. 1944. Ontogenetic change in antigenic specificity of the organs of the chick. *J. Inf. Dis. 74*, 225-233.

Canat, E. H., and E. L. Opie. 1943. Inflammation in embryonic life. I. Changes produced by particulate matter and by a chemical agent. *Am. J. Path. 19*, 371-394.

Chernoff, A. I. 1952. Immunologic studies of fetal hemoglobin. *Science 116*, 518.

Common, R. H., W. A. Rutledge, and W. Bolton. 1947. Influence of gonadal hormones on serum riboflavin and certain other properties of blood and tissues in domestic fowl. *J. Endocr. 5*, 121-130.

Cooper, R. S. 1948. A study of frog egg antigens with serum-like reactive groups. *J. Exp. Zool. 107*, 397-438.

Culbertson, J. T. 1938. Natural transmission of immunity against *Trypanosoma lewisi* from mother rats to their offspring. *J. Parasit. 24*, 65-82.

Darrow, R. R., S. Novakovsky, and M. A. Austin. 1940. Specificity of fetal and of adult human hemoglobin precipitins. *Arch. Path. 30*, 873-880.

Duran-Reynals, F. 1940. The flocculation of tissue extracts by normal and immune sera of fowls and of other animals. *Yale J. Biol. Med. 12*, 361-398.

Dzierzgowski, S. K. 1901. Ein Beitrag zur Frage der Vererbung der künstlichen diphtherieimmunität. *Zentralbl. f. Bact. I. Abt. 30*, 2263-2265.

Ebert, J. D. 1950. An analysis of the effects of anti-organ sera on the development *in vitro* of the early chick blastoderm. *J. Exp. Zool. 115*, 351-378.

Ehrlich, P. 1892. Ueber Immunität durch Vererbung und Säugung. *Zeits. Hyg. u. Infekt. Krank. 12*, 183-203.

Everett, J. W. 1935. Morphological and physiological studies of the placenta in the albino rat. *J. Exp. Zool. 70*, 243-280.

Fenner, F. 1948. The epizootic behavior of mouse-pox (infectious ectromelia). *Brit. J. Exp. Path. 29*, 69-91.

Flickinger, R. A., and G. W. Nace. 1952. An investigation of proteins during the development of the amphibian embryo. *Exp. Cell Res. 3*, 393-405.

Frazer, D. T., T. H. Jukes, H. D. Branion, and K. C. Halpern. 1934. The inheritance of diphtheria immunity in ducks. *J. Immun. 26*, 437-446.

Goodpasture, E. W., and K. Anderson. 1937. The problem of infection as presented by bacterial invasion of the chorioallantoic membrane of chick embryos. *Am. J. Path. 13*, 149-174.

Grobstein, C. 1950. Production of intra-ocular hemorrhage by mouse trophoblast. *J. Exp. Zool. 114*, 359-374.

Grobstein, C. 1950. Behavior of the mouse embryonic shield in plasma clot culture. *J. Exp. Zool. 115*, 297-315.

Grobstein, C. 1951. Intra-ocular growth and differentiation of mouse embryonic shields implanted directly and following *in vitro* cultivation. *J. Exp. Zool. 116*, 501-526.

Grobstein, C. 1952. Effect of fragmentation of mouse embryonic shields on their differentiative behavior after culturing. *J. Exp. Zool. 120*, 437-456.

Grobstein, C. 1952. Intra-ocular growth and differentiation of clusters of mouse embryonic shields cultured with and without primitive endoderm and in the presence of possible inductors. *J. Exp. Zool. 119*, 355-380.

Grosser, O. 1909. Die Wege der fetalen Ernährung. *Samml. anat. physiol. Vortr. 3*, 79-96.

Grosser, O. 1927. *Frühentwicklung, Eihautbildung und Placentation des Menschen und der Saügetiere*. Bergmann, Munchen.

Hall, F. G. 1934-35. Haemoglobin function in the developing chick embryo. *J. Physiol. 83*, 222-228.

Haurowitz, F. 1950. *Chemistry and Biology of Proteins*. Academic Press, New York.

Heine, F. 1936. Phagozytose Versuche am Hühnerembryo. *Arch. f. Entw.-mech. 134*, 283-293.

Hill, R. 1931. A method for the estimation of iron in biological materials. *Proc. Roy. Soc. (B) 107*, 205-214.

Hoff-Jorgensen, E., and E. Zeuthen. 1952. Evidence of cytoplasmic desoxyribosides in the frog's egg. *Nature 169*, 245.

Holtfreter, J. 1946. Experiments on the formed inclusions of the amphibian egg. I. The effect of pH and electrolytes on yolk and lipochondria. *J. Exp. Zool. 101*, 355-405.

Irwin, M. R. 1949. Immunological studies in embryology and genetics. *Quart. Rev. Biol. 24,* 109-123.

Irwin, M. R. 1951. Genetics and immunology, in Dunn, *Genetics in the 20th Century.* Macmillan.

Jukes, T. H., D. T. Frazer, and M. D. Orr. 1934. The transmission of diphtheria antitoxin from hen to egg. *J. Immun. 25-26,* 353-360.

Keeler, C. E., and W. E. Castle. 1933. A further study of blood groups of the rabbit. *Proc. Nat. Acad. Sci. 19,* 403-411.

Keeler, C. E., and W. E. Castle. 1934. Blood group inheritance in rabbits. *J. Hered. 25,* 433-439.

Kemp, T. 1930. Über die Empfindlichkeit der Blutkörperchen gegenüber isohämagglutininen im Fötalleben und im Kindesalter beim Menschen. *Acta Path. Microbiol. Scand. 7,* 146-156.

Kendrew, J. C. 1949. Foetal hemoglobin. *Endeavour 8,* 80-85.

Klemperer, F. 1893. Über natürliche Immunität und ihre Verwertung für die Immunizierungstherapie. *Arch. f. Exper. Path. u. Pharmac. 31,* 356-382.

Kolodny, M. H. 1939. The transmission of immunity in experimental trypanosomiasis (*Trypanosoma cruzi*) from mother rats to their offspring. *Am. J. Hyg. 30,* 19-39.

Kuttner, A., and B. Ratner. 1923. The importance of colostrum to the newborn infant. *Am. J. Dis. Child. 25,* 413-434.

Levene, P. 1948. The mechanism of transplacental immunization. *Blood 3,* 404-413.

Marshall, M. E., and H. F. Deutsch. 1950. Some protein changes in fluids of the developing chick embryo. *J. Biol. Chem. 185,* 155-161.

Marshall, M. E., and H. F. Deutsch. 1951. Distribution of egg-white proteins in chicken serum and in egg yolk. *J. Biol. Chem. 189,* 1-9.

McCance, R. A., A. O. Hutchinson, R. F. A. Dean, and P. E. H. Jones. 1949. The cholinesterase activity of the serum of newborn animals and of colostrum. *Biochem. J. 45,* 493-496.

McCarthy, M., and O. T. Avery. 1946. Studies on the chemical nature of the substance inducing transformation of Pneumococcal types. III. Effect of desoxyribonuclease on the biological activity of the transforming substance. *J. Exp. Med. 83,* 89-96.

Miller, W. J. 1953. The time of appearance of species-specific antigens of *Columba guinea* in the embryos of backcross hybrids. *Physiol. Zool. 26,* 124-130.

Mossman, H. W. 1926. The rabbit placenta and the problem of placental transmission. *Am. J. Anat. 37,* 433-497.

Nace, G. W. 1953. Serological studies of the blood of the developing chick embryo. *J. Exp. Zool. 122,* 423-448.

Nace, G. W., and A. M. Schechtman. 1948. Development of non-vitelloid substances in the blood of the chick embryo. *J. Exp. Zool. 108,* 217-234.

Nicholas, J. S. 1937. The development of rat embryos in a circulating medium. *Anat. Rec. 70,* 199-210.

Nicholas, J. S. 1942. Experiments on developing rats. IV. The growth and differentiation of eggs and egg-cylinders when transplanted under the kidney capsule. *J. Exp. Zool. 90*, 41-71.

Nicholas, J. S. 1947. Experimental approaches to problems of early development in the rat. *Quart. Rev. Biol. 22*, 179-195.

Nicholas, J. S., and D. Rudnick. 1933. The development of rat tissues upon the chick chorioallantois. *J. Exp. Zool. 66*, 193-261.

Painter, T. S. 1940. On the synthesis of cleavage chromosomes. *Proc. Nat. Acad. Sci. 26*, 95-100.

Pedersen, K. O. 1944. Fetuin, a new globulin isolated from ralf serum. *The Svedberg*, 490-499.

Pedersen, K. O. 1947. Ultracentrifugal and electrophoretic studies on fetuin. *J. Phys. Colloid. Chem. 51*, 164-171.

Piccarillo, R., and A. M. Schechtman. Unpublished.

Pickering, J. W., and R. J. Gladstone. 1925. The development of blood plasma. I. The genesis of the coagulable material in embryo chicks. *Proc. Roy. Soc. (B) 98*, 516-522.

Ramsay, W. N. M. 1951. Iron metabolism and haemoglobin formation in the embryonated hen egg. 2. Some observations on the embryonic blood supply. *Biochem. J. 49*, 494-498.

Romanoff, A. L., and A. J. Romanoff. 1949. *The Avian Egg*. John Wiley & Sons, New York.

Roughton, F. J. W., and J. C. Kendrew. 1949. *Haemoglobin*. Interscience Publishers, New York.

Rudnick, D. 1938. Differentiation in culture of pieces of the early chick blastoderm. I. The definitive primitive streak and head-process stages. *Anat. Rec. 70*, 351-368.

Rywosch, M. 1907. Ueber Hämolyse und Bactericidie des embryonalen Hühnerblutes. *Zentralbl. f. Bakt. u. Infekt. krank. I. Abt. 44*, 468-474.

Sabin, F. R. 1920. Studies on the origin of blood vessels and of red blood cells as seen in the living blastoderm of chicks during the second day of incubation. *Carnegie Contrib. to Embry. 9*, 213-262.

Sabin, F. R. 1939. Cellular reactions to a dye-protein with a concept of the mechanism of antibody formation. *J. Exp. Med. 70*, 67-82.

Sachs, H. 1903. Ueber Differenzen der Blutbeschaffenheit in verschiedenen Lebensalter. *Zentralbl. f. Bakt. I. Abt. 34*, 686-692.

Schechtman, A. M. 1952. Physical and chemical changes in the circulating blood. *Am. N.Y. Acad. Sci. 55*, 85-98.

Schechtman, A. M., and H. Hoffman. 1952. Serological studies on the origin of globulins in the serum of the chick embryo. *J. Exp. Zool. 120*, 375-390.

Schechtman, A. M., G. W. Nace, and T. Nishihara. Unpublished.

Schjeide, O. A., and A. M. Schechtman. Unpublished.

Schultz, J. 1952. Interrelationships between nucleus and cytoplasm and problems at the biological level. *Exp. Cell Res. Suppl. 2*, 17-42.

Shepard, C. C., and G. A. Hottle. 1949. Studies of composition of livitin fraction of yolk in hen's eggs. (Use of electrophoretic analysis.) *J. Biol. Chem. 179*, 349-357.

Sherman, H. W. 1919. Antibodies in the chick. *J. Inf. Dis. 25,* 256-261.

Shope, R. E., and H. J. Griffiths. Rinderpest. VI. The persistence of virus in chicks hatched from infected eggs. *Am. J. Vet. Res. 7, Part 2,* 170-174.

Smith, C. A. 1951. *The Physiology of the Newborn Infant.* Charles C. Thomas, Springfield, Ill.

Sölling, P. 1937. Der Komplementgehalt von Seren Neugeborenen, Säuglingen und Früchten. *Zeits. f. Immun. forsch. 91,* 15-21.

Sonneborn, T. M. 1947. Developmental mechanisms in paramecium. *Growth 11,* 291-307.

Spratt, N. T. 1948. Development of the early chick blastoderm on synthetic media. *J. Exp. Zool. 107,* 39-64.

Steinmuller, O. 1937. Phagozytoseversuche am Lebenden Hühnerembryo. *Arch. f. Entw.-mech. 137,* 13-24.

Taylor, K. M., and A. M. Schechtman. 1949. *In vitro* development of the early chick embryo in the absence of small organic molecules. *J. Exp. Zool. 111,* 227-254.

Thomsen, O., and K. Kettel. 1929. Die Starke der menschlichen Isoagglutininen und entsprechenden Blutkörperchenrezeptoren in verschiedenen Lebensaltern. *Zeits. f. Immun. forsch. 63,* 67-93.

Tyler, A. 1947. An auto-antibody concept of cell structure, growth and differentiation. *Growth Suppl. 10,* 7-19.

Waterman, A. J. 1936. Heteroplastic transplantations of embryonic tissues of rabbit and rat. *Am. J. Anat. 60,* 1-25.

Weiss, P. 1947. The problem of specificity in growth and development. *Yale J. Biol. Med. 19,* 236-278.

Weiss, P. 1950. Perspectives in the field of morphogenesis. *Quart. Rev. Biol. 25,* 177-198.

White, A., and T. F. Dougherty. 1946. The role of lymphocytes in normal and immune globulin production and the mode of release of globulin from lymphocytes. *Ann. N.Y. Acad. Sci. 46,* 859-880.

Wiener, A. S. 1943. *Blood Groups and Blood Transfusion.* Charles C. Thomas, Springfield, Ill.

Wiener, A. S. 1946. Recent developments in knowledge of the Rh-Hr blood types; tests for Rh sensitization. *Am. J. Clin. Path. 16,* 477-497.

Wimsatt, W. A. 1949. Cytochemical observations on the fetal membranes and placenta of the bat, *Myotis lucifugus lucifugus. Am. J. Anat. 84,* 63-129.

Windle, W. F. 1940. *Physiology of the Fetus.* W. B. Saunders Co., Philadelphia.

Wolfe, H. R., and E. Dilks. 1948. Precipitin production in the chicken. III. Variation in antibody response as correlated with age of animal. *J. Immun. 58,* 245-250.

Zeuthen, E. 1951. Segmentation, nuclear growth and cytoplasmic storage in eggs of echinoderms and amphibia. *Pub. Staz. Zool. Napoli 23,* 47-69.

II. IMMUNOBIOLOGICAL APPROACH TO SOME PROBLEMS OF INDUCTION AND DIFFERENTIATION

BY M. W. WOERDEMAN[1]

As USED IN EMBRYOLOGY the term induction was originally applied to effects exerted by certain groups of embryonic cells on adjacent groups of less-differentiated cells, with the result that the trend of differentiation of the latter becomes determined. Continued research, however, has revealed that the principle of induction probably applies to many processes concerned with organization and tissue differentiation. Therefore I propose to examine the problems of induction and differentiation in connection with each other.

We owe the discovery of embryonic induction to Spemann. He had found from an extensive series of experiments that the different parts of young amphibian embryos before gastrulation possess much regulatory ability and that consequently no strict determination exists. This state of affairs changes during gastrulation and Spemann was led to the conclusion that the dorsal lip of the blastopore was the center from which organization starts. It is now thirty years ago that, in his laboratory, fragments of the dorsal lip of the blastopore of young *Triton* gastrulae were grafted into hosts of the same age and the phenomenon of embryonic induction was discovered (Spemann and Hilde Mangold, 1924; Spemann, 1924). The grafts not only continued their own development but also exercised an influence upon adjacent groups of cells so that these cells were partly assimilated into the organs formed by the graft and partly formed themselves complementary organs which they would not have formed in the normal course of development. Spemann used the term "induction" for this activity of the graft and adopted the word "organizer" for the inducing graft itself.

It seems highly questionable that the adage "simplicity is a characteristic of truth" holds good for the solution of the ever-increasing problems associated with embryonic induction. I shall not try to sketch in a historical survey their development, but I wish to touch slightly on some aspects of the complicated questions which have arisen as a result of continued research. The different parts of the amphibian gastrula do not reach a definite histological determination synchronously. At a cer-

[1] Anatomisch-Embryologisch Laboratorium, Universiteit van Amsterdam.

[33]

tain phase of development some cells are still interchangeable and capable of regulation whereas other parts possess the capability of self-organization into more or less harmoniously-formed structures. Differential determination of tissues takes place gradually and different grades of determination have been distinguished. Weiss and others properly hold that usage of the term "determination" causes special difficulties.

Stepwise determination is connected with the way in which induction occurs. The primary inductor, the dorsal lip of the blastopore, invaginates and forms the roof of the archenteron, which induces the neural plate and neural crest. In their turn these structures become inductive themselves and provide a series of so-called secondary inductors. In the same way in subsequent stages tertiary and even higher-level inductors arise. Thus it becomes highly probable that embryonic development involves a chain of induction processes and depends on the activities of a great number of different inductors which come into operation in different stages of development. Moreover there are many arguments to support the view that in most cases of induction of a certain organ or tissue development is not due to the activity of a single inductor but to different inductors which come into action successively. Simultaneous activity of some of them is not to be excluded. In this connection Holtfreter (1935a, 1935b) has spoken of coordinated inductor systems.

Although a single inductor may have a strong tendency to induce a special structure, under experimental conditions it may induce other structures as well. So, it is an acceptable conclusion that under normal conditions a single inductor may belong to different inductor systems and its activity may be reinforced or inhibited by other inductors.

While these observations on the way in which inductors operate have given us a better understanding of the induction process, they do not alter the fact that for most induction processes the inductors constituting the coordinated inductor system are not all known; especially beyond our knowledge still is the intricate manner in which they are coordinated. In the search for a clue to the nature of the inductive agents it was found that not only are groups of living embryonic cells capable of induction but so also are killed inductors or extracts made from them. Even living or killed organ fragments from adult animals possess the capability of induction. These remarkable discoveries turned the attention of investigators to the role played by the reacting material. It must be asked to what extent the reacting system determines the result of an induction. Its response is first of all dependent on the reactivity potentialities which it possesses at the moment of induction. These potentialities have been termed the competence of the reacting system (Waddington, 1932). This competence not only decreases with age but also changes qualitatively. This is certainly connected with chemical differentiation and seems to occur not only when the cells remain in their natural surroundings but also in explants. Therefore it is often claimed to be an autonomous process. In my opinion, however, the possibility can not be excluded that it was started in an early stage of development by an inductive action.

In larger cell territories such as the ectoderm change of competence does not take place everywhere at the same time or at the same rate; thus regional differences of competence are set up. Further the rate of change of competence seems to differ greatly in various species of animals, thus introducing a genetical factor. Also it has been found that the effect of an induction is under the control of the genetic potentialities of the reacting system whereas inductors in general do not seem to be species-specific. Thus heteroplastic and xenoplastic transplantation experiments have shown that the eye cup of an anuran amphibian can induce a lens in the ectoderm of a urodele embryo and vice versa, but the lens will show the characteristics of the host species. Spemann and Schotté (1932) grafted anuran trunk ectoderm into the prospective mouth region of newts. The urodele host induced oral structures in the anuran ectoderm, but they were typical for anuran amphibians: horny teeth and suckers. In the opposite case an anuran host induced dentine teeth and balancers in newt ectoderm.

All these observations on the contribution of the reacting system to the results of an induction process oblige us to reexamine the role of the inductors. Thus far I have avoided the use of the term organizer and have spoken of inductors although the word organizer is still frequently used. However, when a killed fragment of adult tissue or a chemical substance induces a more or less harmoniously built structure, the inductive agent can hardly be called an organizer. The organization of the induced structure must be due to the particularities of the reacting system. In the production of horny teeth and suckers by the oral inductors of a urodele embryo the organization of these organs must also be ascribed to the properties of the reacting ectoderm. In the first described case of embryonic induction by fragments of the dorsal lip of the blastopore the inductor induced structures which became incorporated into and assimilated with the derivatives of the inductor itself besides other structures which remained independent of these derivatives. Together they formed an harmoniously built part of a so-called secondary embryo. Hence it seems that in this case Spemann's term organizer could be defended. In nearly all other cases of induction, however, the inductors do not deserve the name of organizers and the induced structures seem to be to a large extent self-organized. When I say "to a large extent," it is because I am well aware of the fact that there are observations which indicate that the structure of some induced organs is dependent on the organ pattern of the inductor. However, the situation is perplexing in that well-organized structures can develop under the inductive activity of atypical inductors such as extracts from inductors. Furthermore, Holtfreter (1944) and Deuchar (1953) have shown that fragments of the archenteric roof can apparently induce perfectly organized head, trunk, and tail structures after its cells have been first disaggregated in alkali and then allowed to reaggregate in random order. No wonder that in consideration of the vast amount of conflicting evidence no uniform opinion can be found among embryologists.

Waddington (1940) and Needham (1942) distinguish between evocation and individuation. Are the inductors merely evocators of a certain cell type? Do they determine only the cytological characteristics of the induced structures, or also their organizational pattern? Or does the individuation (formation of the organ pattern) depend on the activities of the reacting material? Extreme conceptions ascribe an even less important role to the inductors and regard them as mere activating mechanisms. These questions keep embryologists divided at the present time. The answers are largely connected with the conceptions of the nature of the inducing agent. From experiments with killed inductors and extracts it seems probable that inductive activity is of a chemical nature. Various efforts have been made to isolate from killed inductors and their extracts the active inducing substances, and several substances have been claimed to be inductive agents. Originally some embryologists cherished the hope that a single chemical factor of induction might be found, but it soon appeared that they had oversimplified the problem. There came a time when the discovery of more and more induction processes gave rise to the conception of specific inductive substances, each of them responsible for the development of specific structures. The inductors would have to be qualitatively different. However, although a single inductor may show a certain specific activity, we must bear in mind that many inductions occur by the action of inductor systems, and an inductor may belong to several such inductor systems. In the above-mentioned experiments of Spemann and Schotté the inductors exhibit a regional specificity inasmuch as they induce a complex of oral structures, but they can hardly be termed specific inductors for specific organs. The inductor for balancers can apparently induce suckers also.

It seems correct to assume the presence of qualitatively different substances for some examples of induction processes. But the question of whether not only qualitative differences but also quantitative differences are responsible for some inductive activities is still open to discussion. The assumption of the presence of a small number (one or two) of hypothetical substances in gradients of concentration, and thresholds for their action which will account for all kinds of inductions (Dalcq; Pasteels), may have had value as a working hypothesis, but in my opinion it fails to explain the great complexity of the induction processes. Also I fear that neither the hypothetical evocators and modulators of Waddington (1940) nor the eidogens of Needham (1942) can give a satisfactory explanation of induction.

Finally, an hypothesis of Holtfreter (1948) should be mentioned. Stressing the observation that many artificial inductors may have had a cytolytic effect on the cells of the reacting system, he suggests that in the induction of neural plates the inductors may cause a mild cytolysis of the ectodermal cells and in so doing set free neuralizing factors which start a series of processes leading to the formation of neural cells. The inductive agent may be unspecific if it causes only the change in the cell surface which is necessary for the release of neuralizing factors (neurogenes). The old controversy, however, is again raised. Eakin (1949) is of the opinion that one can not exclude the possibility that neuralizing factors originate from the inductor (which then should be a specific neural inductor) and penetrate into the ectodermal cells, thus enabling them to show neural differentiation.

During the past thirty years of research the problem of embryonic induction has become more and more complicated. Every new approach in the field is still welcomed by embryologists. One new development is the expectation of additional data from biochemical investigations. The discovery that glycogen disappears from the cells of the dorsal lip of the blastopore during gastrulation (Woerdeman, 1933a, 1933b) has led to an extensive study of the metabolic processes in the inductors and other parts of the embryo. Histochemistry has also yielded valuable data concerning the presence of certain compounds in embryonic cells. Such work, however, needs corroboration through the application of other techniques. The structural differences among cells must depend upon the component materials of their protoplasm, among which proteins are of paramount significance. The regulation of the synthesis of these components is dependent on the presence of enzymes, and specific differentiations of cells must be explained by differences in their enzyme equipment and the production of specific proteins. The biochemistry of the enzymes therefore has become a very important tool in the study of induction and differentiation.

A second new approach from which inductive and differentiative processes may be studied is the detection of specific proteins by immunological methods. This field of research is still young but promises to lead to valuable results and to the solution of some perplexing questions. I will mention only a few. First of all the question can be raised as to whether the various specific proteins known to exist in the adult are already present in the egg. If this question be answered in the negative, another arises, namely at what time in embryonic development do they make their appearance? Further, it may be that there will be transitional antigens in the embryo which disappear in later stages. What is the role they play in embryonic development? It will also be interesting to determine whether specific antigens appear prior to the formation of organ primordia or not and whether a relationship can be detected with induction processes. Also it must be asked whether or not these proteins are essential in the processes of differentiation, that is to say, in the production of a specific cytoplasm.

As far as I know, immunological techniques in the study of embryology were used for the first time as early as 1904 and 1906 by Graham Smith and Braus. These pioneers in the field and a small number of followers prepared immune sera against adult tissues or against larvae and tested saline extracts of eggs and larvae with these antisera. Some of the early workers (Smith, 1904; Braus, 1906; Dunbar, 1910; Uhlenhuth and Haendel, 1910; Uhlenhuth and Wurm, 1939; Kritchevsky, 1914) came to the conclusion that a difference exists between the protein components of the eggs, the larvae, and the adult forms of amphibia and fish. Other investigators, such as Iwae (1915), Guggenheim (1929), and Witebsky and Szepsenwol (1934), studying Forssman's antigen during the development of the chick, asserted that it can already be found in the egg. Rössle (1905) and Wilkoewitz and Ziegenspeck (1928) found no differences in antigenicity between the embryonic and adult organism. Kritchevsky (1923) and Abe (1931) arrived at the conclusion that some antigens appear only during embryonic development and that their quantity increases regularly. Most of this work, however, was done with techniques which included some rather serious errors, hence the results must be judged cautiously.

Immunobiological techniques are continually being improved since it has been shown that they require very critical application and careful handling in order to avoid erroneous results. Many proteins are unstable and become denaturated if the organ extracts are not prepared with extreme precaution. Also organ extracts contain not only the specific antigens but also antigens from tissues common to almost all other organs, such as blood, connective tissue, etc. Antisera prepared with such organ extracts must be absorbed with those common antigens in order to purify them and improve their specificity. Still, organ specificity is difficult to obtain and often quantitative tests have to be made after determining the titer of the antiserum (which is the highest dilution in which it is capable of reacting, in the precipitin test, with a standard solution of antigen).

Modern work, taking into account the various sources of error of immunological methods, has been devoted to the detection of the specific antigens of the blood in the embryo. Since the results of these studies have been dealt with in Chapter I, I shall restrict myself to a brief mention of Schechtman and his collaborators Flickinger and Nace (Schechtman, 1947, 1948; Schechtman and Nace, 1950; Flickinger and Nace, 1952), and also Moore, Shen, and Alexander (1945), Cooper (1946, 1948, 1950) and Spar (1951), all of whom contributed to our knowledge of antigens of blood in embryonic stages. Regarding tissue specific proteins, only limited data are yet available. Most of them bear upon the development of the chick. Here again Schechtman should be mentioned together with Burke and his collaborators (1944) and Ebert (1950, 1951a, 1951b, 1952).

Schechtman (1948) utilized saline extracts of brain, heart, liver, and muscle from chick embryos 19 to 20 days old as antigens and tested extracts of primitive streak and neurula stages. The latter contained common antigens different from those of both yolk and blood and from the Forssman antigen. Extracts of the early chick embryos gave a positive precipitin test with antibrain serum absorbed with blood. Absorption with brain extract or extracts of heart and liver removes all precipitins. The early stages thus seem to contain a general organ antigen which is not specific for any one of the organs mentioned.

Ebert repeated Schechtman's investigations and obtained similar results with definitive primitive streak and early somite stages. In one respect only did his results differ. His antisera absorbed with extracts

[39]

of adult liver did not remove precipitins for either the adult tissue specific antigens or the embryonic antigens. This divergence in results is
probably explained as the result of the different methods employed by
Schechtman and Ebert in preparing the antigen solutions.

Ebert, well aware of the difficulties caused by the low grade of purity
of the antigens, has tried to use a highly purified protein as antigen.
Muscle actin and myosin were not found to be very satisfactory. Therefore he turned to melanin and, together with Goodgal, combined it with
a protein to reach a high precipitin titer. They have succeeded in obtaining a serum containing potent antibodies (Ebert and Goodgal, 1949).

In Ebert's paper published in 1952 he mentions that studies are now
in progress with black and red melanins from embryonic sources. As
the appearance of melanin pigments in embryonic differentiation is a
striking criterion of differentiation which can be easily detected in microscopical sections, a combination of morphological and immunobiological
investigations of pigmentation promises valuable information. In the
same paper Ebert presented the preliminary results of an investigation
of the ontogenetic changes in the antigenic specificity of the chick
spleen. The serum-agar technique of Oudin (1946, 1947) as modified
by Munoz and Becker (1950) was closely followed. In the spleen of
the 12-day chick at least three separate antigens could be detected. In
the period between the twelfth and eighteenth days at least three new antigens make their appearance. Attempts were made at purification of these
antigens, and Ebert makes it clear that a combination of the serum-agar
technique with chemical analysis offers great promise as a critical tool
for investigations into the question of time and manner of origin of
specific proteins.

Burke, Sullivan, Petersen, and Weed (1944) demonstrated that adult
antigens of erythrocytes, lens, kidney, brain, and gonad appear in the
chick embryo at various hours of incubation and in the sequence mentioned. They also showed that lens and brain antigens change during
development. For example antigens of young stages were not reactive
with antisera to older stages. They believe that they have found at least
two different antigens in the developing brain. A further interesting approach to the study of the antigenic properties of young embryos was
their effort to influence the development of the chick lens by placing
antilens serum on the chorioallantoic membrane. Antiserum to adult lens
caused partial lysis of the lens only in embryos of from 146 to 192 hours.

Similar experiments were made by Grunwaldt, Ebert, and Clayton.
Grunwaldt (1949) examined the effects of antisera against three frac-

tions of newly hatched chicken brain on cultures of 9-day and 13-day chick embryo spinal cord and came to the conclusion that an alcohol insoluble antigen appears about the ninth day and an alcohol soluble one between the ninth and thirteenth day of incubation. Ebert (1950, 1951a) cultivated chick blastoderms (primitive streak and head process stages) on a medium containing various concentrations of rabbit antisera to adult chicken heart, spleen, or brain. Control cultures were made with normal rabbit serum and a wide range of concentrations was tested. The author concludes that antigenic substances identical or closely related to those of the adult organs of the chick must be present in the early chick blastoderm prior to the appearance of the corresponding organ primordia. Evidence is presented that these antigens are localized in fairly definite areas of the blastodisc as early as the primitive streak stage.

The preceding investigations were all carried out on chick material. In amphibian development data are scarce. Clayton (1951) reported on experiments with material of *Triton alpestris*. Young stages of this urodele, reared *in toto* or as fragments, were cultivated as explants in solutions of normal rabbit serum, and of rabbit antiserum to isolated embryonic parts of *Triton*. A large percentage of explants died within the first two days. Morulae explanted into antigastrula serum showed inhibition of gastrulation. This observation is explained by the assumption that at the beginning of or during gastrulation a specific antigen or specific antigens appear. Explants from archenteron and ectoderm of the late gastrula cultured in both antiarchenteron and antiectoderm serum showed anomalies which suggest the possibility that archenteron and ectoderm contain one or more antigens specific to each.[2] It may be expected that more experiments on amphibian material will yield valuable

[2] Some days after the presentation of this paper at the Twelfth Growth Symposium a more extensive publication of the work of Clayton appeared (*J. Emb. Exp. Morph. I*, 1, 25, 1953). In *Triturus* embryos antigenic substances have been synthesized between the blastula and gastrula stages, before neurulation, and between neurulation and tail-bud formation. Six antigenic fractions were found in the gastrula, three of which are to be found only after gastrulation. Ectoderm and archenteron roof contain fractions specific to themselves, but they also contain some antigens in common. Ectoderm and neural plate have some specificity relative to each other.

An antigenic substance, specific to a given tissue type, may be detected before the first appearance of the tissue. Two antigenetic fractions later characteristic of distinct tissues may coexist in a common precursor.

With regard to the application of antisera to cultures of embryonic material, it is stated that partially absorbed antisera tend to cause abnormalities rather than death of the whole embryos. In general, however, antisera exhibit an irregular cytotoxic effect. There are indications that a certain degree of specificity in action of absorbed antisera occurs when used in culture media. A difference in susceptibility of blastulae and gastrulae to antisera which can not be attributed to their difference in antigenic constitution must depend on a change in the location of certain antigens or a change in permeability.

information, especially in connection with the processes of induction, for the very reason that these processes are best known in amphibia.

Although experimental embryology has extensively utilized sea urchin eggs, immunological work on this material is only beginning. Perlmann and Gustafson (1948) found in plutei of *Paracentrotus lividus* a substance which was absent, or at least undetectable, in younger stages. It is probably produced only after blastulae have started gastrulation. This work is being continued in the Wenner-Grens Institute for Experimental Biology in Stockholm, where efforts have also been made to detect maternal and paternal proteins in sea urchin hybrids.

Work done in my laboratory shows that by use of a sensitive method one can study precipitin reactions with small quantities of material. Ten Cate (1949) and ten Cate and Van Doorenmaalen (1950) have been able to investigate extracts of small fragments of head ectoderm of chick and frog embryos with antisera to adult chick and frog lens. The lens of the eye has great advantages for serological work because it is a purely epithelial organ isolated from its surroundings by a capsule and devoid of blood vessels. It is known to contain some highly organ specific proteins. Specific antisera are therefore rather easily prepared.

My collaborators have succeeded in demonstrating the presence of lens proteins in a much earlier stage of development than did Burke. Chick embryos at 60 hours of incubation and frog embryos (*Rana esculenta*) in stage 19/20 (Shumway, 1940) were the youngest embryos in which adult lens proteins could be detected in head ectoderm. Upon microscopical examination it appeared that the lens of a chick embryo at this stage is still an open groove and that the frog embryo in stage 19/20 possesses a solid knot of cells still in broad contact with the ectoderm. Specific histological differentiation occurs only much later. From these observations it seems permissible to conclude that very young lens primordia already contain specific lens protein before microscopically visible histological differentiation sets in.

Ten Cate and other collaborators have tried to isolate the various proteins of the adult lens by biochemical methods and by electrophoresis. A purification of the various fractions is difficult, but they now have in hand an almost pure α-crystalline fraction. Antiserum against this protein seems to be more specific than antisera against the whole lens. We have also investigated whether the various lens proteins appear at different phases of lens development and if there are quantitative differences. Immunological methods [precipitin reactions, also with the agar-serum techniques of Oudin (1947) and Ouchterlony (1948, 1950)], bio-chemical methods, and electrophoresis have been used. Preliminary results of Zuidweg [3] indicate that in chick embryos on the fourth and fifth day of incubation several lens proteins are already present. From the fifth day on there is a regular increase in the protein content of the lens primordia. The a-crystalline fraction increases especially in the period between the fifth and eighth day. From then on it increases more slowly. About the seventh day the other fractions begin to increase rapidly. So, between the sixth and seventh day a change takes place in the production of lens proteins. When we compare these results with the microscopical picture of the lens in this period, we see that this metabolic change seems to precede the formation of a new generation of lens fibers by the so-called marginal zone of the lens epithelium, which starts on the eighth day. We hope to continue these investigations and shall look for a correlation between the results of our protein determinations and the morphological differentiation of the lens.

Another approach to the investigation of lens development is being made by Van Doorenmaalen, who will try to detect in sections of embryos the presence of antigens by combining the γ-globulin fraction of antisera with a fluorescent dye and applying this compound to sections according to a method described by Coons and his co-workers (1942; 1950a,b; 1951; see also Kaplan, 1950, and Hill, 1950). This work is still in progress and to date the results can not be determined.

I myself have tried to study lens induction utilizing the immunological technique developed by ten Cate and Van Doorenmaalen. For this purpose saline extracts were prepared of the head ectoderm of young axolotl neurulae before the appearance of lens placodes. They were not reactive with antiserum to adult lens. The same was found to be the case with extracts of young isolated eye vesicles. A mixture of equal parts of ectoderm extract and extract of eye vesicles, however, showed a positive precipitin reaction after incubation at 37°C. during 24 hours whereas the separate extracts incubated for the same length of time gave negative reactions (Woerdeman, 1950). This work with chick material continues, using trunk ectoderm instead of head ectoderm, but up to the present time no success can be reported.

Our attention has also been turned to lens regeneration. The question arose as to whether the dorsal rim of the iris in urodeles, which can regenerate a lens after extirpation of the original lens, contains lens pro-

[3] Recently published in *Proc. Kon. Ned. Akad. Wetensch. Amsterdam.* Series C. 57, 115. 1954.

teins or possesses the capacity to form them. Van Doorenmaalen, who has begun studying this problem with the Coons technique, has encountered great difficulty in preparing a lens antiserum which is not reactive with the iris. Therefore it will be necessary to study the relationship between the antigens of lens and iris. Most probably substances are to be found in the iris which are closely related to lens proteins. Tissue cultures of iris epithelium of chick embryos of 10 to 12 days, growing on media prepared with extracts of embryos from which the lenses were removed, and which consequently were not reactive with antisera to lens, were extracted with saline. In many cases these extracts gave positive precipitin reactions with antisera which had not given positive reactions with the iris material from which the cultures were made (Woerdeman, 1950). It seems probable that embryonic iris epithelium can produce one or more of the lens antigens.

During the past year, R. L. de Haan, a pupil of Schechtman working in our laboratory, has investigated the muscle components in the regenerating limb of axolotl larvae. He prepared myosin and actin from adult muscle tissue after a slightly modified method of Mommaerts and Parrish and of Hasselbach and Schneider. Rabbit antisera to these substances were found to be rather aspecific. However, absorption experiments with various axolotl tissue extracts, including blood, brain, heart, skin, cartilage, and spleen in various combinations, showed that antisera against axolotl myosin or actin, absorbed with a highly concentrated solution of skin and cartilage, are reactive towards muscle but give a negative reaction with skin, cartilage, blood, brain, liver, and spleen extracts. However, they are not specific to the separate proteins actin and myosin. With these antisera the chemical differentiation of regeneration blastemas was investigated. The hind limbs of axolotl larvae were amputated at the knee. After regeneration blastemas became macroscopically visible, they were removed at intervals of 2 to 3 days distal to the point of original amputation. They were ground and extracted in saline, after which the extracts were tested with antiserum. The stumps from which the blastemas had been removed, and also identical blastemas, were fixed for histological control. Muscle proteins were regularly detected in the blastemas from the twenty-ninth day after amputation on. On the twenty-ninth day there can be seen in the blastema aggregations of cells which form strings of separate slightly elongated cells. These strings usually extend from the tips of the stump muscles, which are degenerating distally, to near the tip of the blastema, where they become indistinguishable from the other cells of the blastema. These

elongated cells, except for their arrangement in strings and their elonga-
tion, are in no way visibly different from the rest of the blastema cells.
About three days later the presence of some muscle fibrils and a beginning cross-striation can be detected in these cells. The preliminary conclusion seems to be that in the regenerating limb of the axolotl specific
muscle proteins become detectable a considerable length of time (in this
case, at least 3 days) before true skeletal muscle fibers become histologically distinguishable.

Langman (1950, 1953) has conducted some experiments which have
a bearing on the problems of this paper, but his results can better be taken
up later.

Although the evidence at present is far too scanty to allow for far-
reaching conclusions, I shall endeavor to formulate some general results
of immunobiological approaches to the problems of induction and differentiation.

Induction seems to be essential for many differentiations and it is possible that it should be regarded as the process which starts the formation
of the specific building stones for differential tissue development. The
fundamental question of whether it activates processes already going on
in the reacting system, or starts new processes which would be impossible
without the activity of the inductor, has not been solved up to the present
time.

Some of the organ and tissue specific antigens are already present in
the early stages of development and have been present for a considerable
time prior to the appearance of corresponding organ primordia. Other
antigens appear only later and their time of appearance can be determined
rather accurately. Connection with certain induction processes has not up
to now been mentioned in the literature except for lens development. The
manner of origin of specific antigens, either by synthesis or the breaking
up of compound molecules, also remains to be examined as does their
role in morphogenesis.

Quantitative and qualitative differences in specific antigens during
development have become evident. Some of these changes are most probably instrumental in morphogenesis although as yet the connection is
not understood. Among the qualitative changes may be mentioned those
which are probably caused by the existence of embryo specific antigens of
a transitory nature which disappear during ontogeny after having contributed to the first steps of differentiation.

It may seem premature at this time to speculate on the conception of

induction and differentiation to which continued immunobiological research may lead us. The experimental biologist, however, must form
hypotheses to direct his research. One hypothesis, set up by P. Weiss in
1939, considered still further in a later publication (Weiss, 1947, 1949,
1950), and reiterated by Tyler (1947) states that growth may be seen
as analogous to the mechanism of antibody formation in which an antigen
causes the production of a globulin of specific structure. I shall not take
up the experimental work done to test this hypothesis by Weiss and his
co-workers Wang, Fischer, Andres, and Ferris (see Weiss and Wang,
1941; discussion to my paper of 1953; Weiss 1950), or comparable work
by Danchakoff (1916, 1918), Willier (1924), Pomerat (1949), and
Ebert (1951a), because these investigations are now well known. They
are of importance not only for a better insight into the general problems
of growth but especially into the production of tissue specific proteins
during tissue differentiation.

It is possible that induction processes resulting in differentiation may
find their explanation in the incorporation of protein molecules into the
cells of the reacting system. These may function as templates and impress
a spatial configuration upon the molecules of the cells coming into direct
physical contact with them, or act as models for molds which would
afterwards turn out more of the incorporated molecules [conceptions of
Breinl and Haurowitz (1930), Alexander (1932), Mudd (1932), Pauling (1940), and Weiss (1947)]. However, one can imagine another
way in which induction functions. It is known that bacteria can form so-
called adaptive enzymes which are formed only when the specific substrates of these enzymes are present in the culture medium. Burnet
(1941) and Burnet and Fenner (1948) have assumed the formation of
such adaptive enzymes, capable of self-replication, to account for the
response to the incorporation of a protein molecule in a cell at the site of
synthesis of an antibody. We do not know with certainty, however,
whether adaptive enzymes can be formed by the cells of higher organisms.
Langman (1950, 1953) has expressed the opinion that adaptive enzymes
may possibly be formed in embryonic cells of mammals cultivated in a
medium containing heterologous blood serum. In our laboratory he cultivated fragments of ovary and testis of rabbit embryos 18 to 20 days old
in media prepared with rabbit serum or with cat serum. In the beginning
there is a marked difference in the mode and rate of growth in the two
media, but after two or three transfers it gradually disappears. By careful
serological experiments with many control tests he has been able to
demonstrate that the gonad fragments of rabbit embryos cultivated for

nearly two weeks in a medium with cat serum contain cat proteins in their cells. They must be able to utilize foreign proteins, for at the time of the experiments their growth did not differ from that of fragments cultivated in rabbit serum. Therefore we must conclude that embryonic mammalian tissues can adapt themselves to the nonspecific components of the culture medium.

We know that transplantation of adult tissues in mammals is mostly unsuccessful unless donor and host are the same individual or at least individuals of the same species. However, tumor cells and embryonic tissue can be transplanted into foreign hosts with more success. This is an indication that the adaptability of these latter cells and tissues is greater than that of adult differentiated cells. Such a situation may perhaps be explained by the capacity of forming adaptive enzymes to a new substrate. If embryonic cells possess this capacity, as Langman assumes, and in induction certain protein molecules are incorporated in the cells of the reacting system, the latter may form adaptive enzymes. This would certainly cause a change in their metabolic performance and could explain why, as a result of induction, differentiation becomes visible. The undeniable specificity of some induction processes would find its explanation in the fact that the molecular structure of newly formed molecules in the reacting cells depends on the configuration of the incorporated molecule, according to the conception of Weiss, or on the circumstance that there is a specific relationship between adaptive enzyme and substrate. Nuclear control of protein formation has been discussed by Caspersson (1947) and enzyme formation is most probably also gene-controlled. This would account for the influence of heredity on induction processes.

Finally, inductive stimuli seem to be especially operative if there is a close contact between inductor and reacting system. Billingham and Medawar (1948) have studied the effect of transplanting in the guinea pig an autograft of pigmented skin to a nonpigmented area. Pigmentation was seen to spread into this unpigmented field. Dendritic cells with a negative dopa-reaction became dopa-positive and changed into melanoblasts. The authors exclude a diffusion of pigment or pigment-forming enzymes, but assume a kind of "infection" of the dendritic cells along well-developed cytoplasmic connections. Their conclusion is that the enzyme equipment of a cell can be increased by implantation of a similar tissue with a different enzyme content and that this phenomenon takes place by protoplasmic contact. Willier's work has made us familiar with the complexity of the processes by which pigmentation is achieved.

[47]

I must therefore leave undecided whether in the case of the experiments of Billingham and Medawar the explanation of the authors is justified. That similar processes occur in induction does not seem too fantastic an idea, however.

Great importance must be attached to the spatial arrangement of the enzymes in the cell. The rate of enzymic reactions will depend on this arrangement and that of their substrates within the cell. Seckel (1938), in the case of diastase and glycogen in the liver cell, has found arguments to support the supposition that surface-active agents may change the accessibility of the enzyme to the substrate. In induction, Weiss (1949) attaches great importance to the molecular structure of the outer layer of the cell, and Holtfreter (1948) assumes that inductors may cause a slight cytolysis, thus changing the cell surface and releasing specific substances within the protoplasm. I have myself always advocated the great importance of changes in the cell surface during induction (Woerdeman, 1937, 1938a,b). So, if we consider what is going on during induction, we may suppose that at the site of contact between inductor and reacting system there occurs a series of interactions which bring about changes in the cell surface. They may lead to a spatial rearrangement of enzymes and substrates in the cell, to a release of some molecules, and to the uptake of others. If the latter are protein molecules, they may set up processes comparable to antigen-antibody reactions either by the formation of new protein molecules or by the formation of adaptive enzymes. These changes in the enzyme equipment of the reacting cells and the formation of new proteins will be followed by metabolic changes which finally find their visible expression in differentiation. They must be gene controlled.

Although the ideas just discussed might constitute an acceptable conception of cytological differentiation, they do not account for organization. We have to assume that in the inductor, as well as in the reacting system, there are patterns of distribution of material, of metabolic processes, and arrangements of molecules which interact in the processes which underlie the origin of an organ pattern (Woerdeman, 1937, 1938a,b). These patterns may change during differentiation and new patterns may arise, thus leading to the well-known organizational differentiation at the end of embryonic development. Many of the processes, which I suppose to take place, have some likeness to the antigen-antibody concepts. Therefore, we may be confident that a further approach to the problem of induction and differentiation from the immunobiological angle holds the promise of a concept better founded than the one which I have been able to present at this time.

BIBLIOGRAPHY

Abe, M. 1931. Antigenic properties of the organ lipoids (alcohol extract) of the human fetus and the new-born (I and II). *Japan J. Obstet. Gynec.* *14, 2,* 28.

Alexander, J. 1932. Some intracellular aspects of life and disease. *Protoplasma 14,* 296.

Billingham, R. S., and P. B. Medawar. 1948. Pigment spread and cell heredity. *Heredity 2, 29.*

Braus, H. 1906. Über das biochemische Verhalten von Amphibienlarven. *Arch. f. Entw.-mech. 22,* 564.

Breinl, S., and F. Haurowitz, 1930. Chemische Untersuchung des Präzipitates aus Hämoglobin und Anti-Hämoglobin-Serum und Bemerkungen über die Natur der Antikörper. *Zeits. f. physiol. Chem. 192,* 45.

Burke, V., N. P. Sullivan, H. Petersen, and R. Weed. 1944. Ontogenetic change in antigenic specificity of the organs of the chick. *J. Inf. Dis. 74,* 225.

Burnet, F. M. 1941. The production of antibodies. *Walter and Eliza Hall Inst. Monogr.* No. 1, Macmillan, Melbourne.

Burnet, F. M., and F. Fenner. 1948. Genetics and immunology. *Heredity 2,* 289.

Caspersson, T. 1947. The relation between nucleic acid and protein synthesis. *Symp. Soc. Exp. Biol. Cambridge 1,* 127.

ten Cate, G. 1949. The presence of adult lens protein in young lens vesicles of chicken and frog embryos. *Natuurwetensch. Tijdschr. Gent.* 92.

ten Cate, G., and W. J. van Doorenmaalen. 1950. Analysis of the development of the eye-lens in chicken and frog embryos by means of the precipitin reaction. *Proc. Kon. Ned. Akad. Wetensch. Amsterdam 53,* 894.

Clayton, R. M. 1951. Antigens in the developing newt embryo. *Nature 168,* 120.

Clayton, R. M. 1953. Distribution of antigens in the developing newt embryo. *J. Embr. and Exp. Morph. 1,* 25.

Coons, L., A. H. Creech, H. J. Jones and E. Berliner. 1942. The demonstration of pneumococcal antigen in tissues by the use of fluorescent antibody. *J. Immun. 45,* 159.

Coons, A. H., and M. H. Kaplan. 1950a. Localization of antigen in tissue cells. *J. Exp. Med. 91,* 1.

Coons, A. H., J. C. Snyder, F. S. Cheever, and E. S. Murray. 1950b. Localization of antigen in tissue cells. IV. Antigens of Rickettsiae and mumps virus. *J. Exp. Med. 91,* 31.

Coons, A. H. 1951. Fluorescent antibodies as histochemical tools. *Fed. Proc. 10,* 558.

Cooper, R. S. 1946. Adult antigens (or specific combining groups) in the egg, embryo and larva of the frog. *J. Exp. Zool. 101,* 143.

Cooper, R. S. 1948. A study of frog egg antigens with serum-like reactive groups. *J. Exp. Zool. 107,* 397.

[49]

Cooper, R. S. 1950. Antigens of frog embryos and of adult frog serum studied by diffusion of antigens into agar columns containing antisera. *J. Exp. Zool. 114,* 403.

Danchakoff, V. 1916. Equivalence of different hematopoietic anlages (by method of stimulation of their stem cells). *Am. J. Anat. 20,* 255.

Danchakoff, V. 1918. Equivalence of different hematopoietic anlages (by method of stimulation of their stem cells). II. Grafts of adult spleen on the allantois and response of the allantoic tissues. *Am. J. Anat. 24,* 127.

Deuchar, E. M. 1953. The regional properties of amphibian organizer tissue after disaggregation of its cells in alkali. *J. Exp. Biol. 30,* 1, 18.

Dunbar, W. P. 1910. Über das serobiologische Verhalten der Geschlechtszellen. II. Mitteilung. *Zeits. f. Immun. forsch. 7,* 454.

Eakin, R. M. 1949. The nature of the organizer. *Science 109,* 2826, 195.

Ebert, J. D. 1950. An analysis of the effects of anti-organ sera on the development, *in vitro,* of the early chick blastoderm. *J. Exp. Zool. 115,* 351.

Ebert, J. D. 1951a. Ontogenetic change in the antigenic specificity of the chick spleen. *Physiol. Zool. 24,* 20.

Ebert, J. D. 1951b. Ontogenetic change in the antigenic specificity of the chick spleen. *Anat. Rec. 111,* 3, 547.

Ebert, J. D. 1952. Appearance of tissue-specific proteins during development. *Ann. N.Y. Acad. Sci. 55,* 67.

Ebert, J. D., and S. H. Goodgal. 1949. On the presence in the sera of domestic fowl of naturally-occurring "agglutinins" for certain melanins. *Anat. Rec. 105,* 141.

Flickinger, R. A., and G. W. Nace. 1952. An investigation of proteins during the development of the amphibian embryo. *Exp. Cell Res. 3,* 393.

Grunwaldt, E. 1949. The evaluation of anti-brain sera by tissue culture methods. *Texas Rep. Biol. Med. 7,* 270.

Guggenheim, A. 1929. Über Antigenfunktionen der Lipoide des Eidotters. *Zeits. f. Immun. forsch. 61,* 361.

Hill, A. G. S., H. W. Deane, and A. H. Coons. 1950. Localization of antigen in tissue cells. V. Capsular polysaccharide of Friedländer bacillus, type B, in the mouse. *J. Exp. Med. 92,* 35.

Holtfreter, J. 1935a. Morphologische Beeinflussung von Urodelenektoderm bei xenoplastischer Transplantation. *Arch. f. Entw.-mech. 133,* 367.

Holtfreter, J. 1935b. Über das Verhalten von Anurenektoderm in Urodelenkeimen. *Arch. f. Entw.-mech. 133,* 427.

Holtfreter, J. 1944. Experimental studies on the development of the pronephros. *Rev. Can. Biol. 3,* 220.

Holtfreter, J. 1948. Concepts on the mechanism of embryonic induction and their relation to parthenogenesis and malignancy. *Growth. Sym. Soc. Exp. Biol. II,* Academic Press, New York.

Irwin, M. R. 1949. Immunological studies in embryology and genetics. *Quart. Rev. Biol. 24,* 109.

Iwae, S. 1915. *Mitteil. med. Gesellsch. Fukuoka, 9,* 1 (cited in Needham, J. 1931. *Chemical Embryology.* Cambridge University Press).

Kaplan, M. H., A. H. Coons, and H. W. Deane. 1950. Localization of antigen in tissue cells. III. Cellular distribution of pneumococcal polysaccharides types II and III in the mouse. *J. Exp. Med. 91,* 15.

Kritchevsky, I. L. 1914. Ein Versuch der Anwendung der Immunitäts-reaktionen für das Studium des biogenetischen Grundgesetzes. *Centralbl. f. Bakt. u. Parasitenk. 72,* 81.

Kritchevsky, I. L. 1923. The relation of immunity reactions to the biogenetic law. Investigations of the chemical structure of the protoplasm of animals during embryonic development by means of heterogeneous hemolysins. *J. Inf. Dis. 32,* 192.

Langman, J. 1950. *Weefselkweek en transplantatie.* Doctor's thesis, Amsterdam.

Langman, J. 1953. Tissue culture and serological experiments. *Proc. Kon. Ned. Akad. Wetensch. Amsterdam, Series C, 56,* 2, 219.

Moore, D. H., S. C. Shen, and C. S. Alexander. 1945. The plasma of developing chick and pig embryos. *Proc. Soc. Exp. Biol. Med. 58,* 307.

Mudd, S. 1932. A hypothetical mechanism of antibody formation. *J. Immun. 23,* 423.

Munoz, J., and E. L. Becker. 1950. Antigen-antibody reactions in agar. I. Complexity of antigen-antibody systems as demonstrated by a serum-agar technique. *J. Immun. 65,* 47.

Nace, G. W. 1953. Serological studies of the blood of the developing chick embryo. *J. Exp. Zool. 122,* 423.

Nace, G. W., and A. M. Schechtman. 1948. Development of non-vitelloid substances in the blood of the chick embryo. *J. Exp. Zool. 108,* 217.

Needham, J. 1931. *Chemical Embryology.* Cambridge University Press.

Needham, J. 1942. *Biochemistry and Morphogenesis.* Cambridge University Press.

Ouchterlony, Ö. 1948. Antigen-antibody reactions in gels. *Arkiv Kemi, Mineral., Geolog. 26B,* 14.

Ouchterlony, Ö. 1950. Antigen-antibody reactions in gels. II. Factors determining the site of the precipitate. III. The time factor. *Arkiv Kemi 1,* 7, 9.

Oudin, J. 1946. Méthode d'analyse immunochimique par précipitation spécifique en milieu gélifié. *Compt. Rend. Acad. Sci. Paris 222,* 115.

Oudin, J. 1947. L'analyse immunochimique du sérum de cheval par précipitation spécifique en milieu gélifié (premiers résultats). *Bull. Soc. Chim. Biol. 29,* 140.

Pauling, L. 1940. A theory of the structure and process of formation of antibodies. *J. Am. Chem. Soc. 62,* 2643.

Perlmann, P., and T. Gustafson. 1948. Antigens in the egg and early developmental stages of the sea urchin. *Experientia 4,* 481.

Pomerat, C. M. 1949. Morphogenetic effects of spleen antigen and antibody administrations to chick embryos. *Exp. Cell Res. Suppl. I,* 578.

Roessle, H. 1905. Über die chemische Individualität der Embryonalzellen. *Münch. med. Wochenschr. 52,* 1276.

Schechtman, A. M. 1947. Antigens of early developmental stages of the chick. *J. Exp. Zool. 105,* 329.

[51]

Schechtman, A. M. 1948. Organ antigen in the early chick embryo. *Proc. Soc. Exp. Biol. Med. 68, 263.*

Schechtman, A. M., and G. W. Nace. 1950. Development of serum proteins in the embryonic chick. *Anat. Rec. 106, 436.*

Seckel, H. P. G. 1938. The influence of various physiological substances on the glycogenolysis of surviving rat liver. *Endocrinology 23, 751.*

Shumway, W. 1940. Stages in the normal development of Rana pipiens. *Anat. Rec. 78, 139.*

Smith, Graham, in G. H. F. Nuttall. 1904. *Blood Immunity and Blood Relationship.* Cambridge University Press.

Spar, I. L. 1951. Antigenic differences among early embryonic stages of Rana pipiens. *Anat. Rec. 111, 451.*

Spemann, H. 1924. Über Organisatoren in der tierischen Entwicklung. *Naturwis. 12, 1092.*

Spemann, H., and H. Mangold. 1924. Über Induktion von Embryonalanlagen durch Implantation artfremder Organisatoren. *Arch. f. mikr. Anat. u. Entw.-mech. 100, 599.*

Spemann, H., and O. Schotté. 1932. Über xenoplastische Transplantation als Mittel zur Analyse der embryonalen Induktion. *Naturwis. 20, 463.*

Tyler, A. 1947. An auto-antibody concept of cell structure, growth and differentiation. *Growth (Suppl.) 10, 7.*

Uhlenhuth, P., and Haendel. 1910. Die Anaphylaxie-Reaktion mit besonderer Berücksichtigung der Versuche zu ihrer praktischen Verwertung. *Ergebn. wissensch. Med. 2, 1.*

Uhlenhuth, P., and K. Wurm. 1939. Über Antikörper gegen Froschlaich. *Zeits. f. Immun. forsch. 96, 183.*

Waddington, C. H. 1932. Experiments on the development of chick and duck embryos, cultivated *in vitro. Phil. Trans. Roy. Soc. London (B) 221, 179.*

Waddington, C. H. 1940. *Organisers and genes.* Cambridge University Press.

Weiss, P. 1939. Size of embryonic organs as detector of organ-specific serological effects. *Anat. Rec. Suppl. 75, 67.*

Weiss, P. 1947. The problem of specificity in growth and development. *Yale Jour. Biol. Med. 19, 235.*

Weiss, P. 1949. The problem of cellular differentiation. *Proc. First Nat. Cancer Con. 50.*

Weiss, P. 1950. Perspectives in the field of morphogenesis. *Quart. Rev. Biol. 25, 177.*

Weiss, P., and H. Wang. 1941. Growth response of the liver of embryonic chick hosts to the incorporation in the area vasculosa of liver and other organ fragments. *Anat. Rec. Suppl. 79, 62.*

Wilkoewitz, K., and H. Ziegenspeck. 1928. Die verschiedenen Generationen und Jugend- und Alters-formen in ihrer Einwirkung auf den Ausfall der Präcipitinreaktionen. *Bot. Arch. 22, 227.*

Willier, B. H. 1924. The endocrine glands and the development of the chick. I. The effects of thyroid grafts. *Am. J. Anat. 33, 67.*

Witebski, E., and J. Szepsenwol. 1934. L'antigène "Forssman" chez les embryons de poulet à différents stades. *Compt. Rend. Soc. Biol. Paris* *115*, 921.

Woerdeman, M. W. 1933a. Über den Glykogenstoffwechsel des Organisationszentrums in der Amphibiengastrula. *Proc. Kon. Ned. Akad. Wetensch. Amsterdam 36*, 2, 189.

Woerdeman, M. W. 1933b. Über den Glykogenstoffwechsel tierischer "Organisatoren." *Proc. Kon. Ned. Akad. Wetensch. Amsterdam 36*, 4, 423.

Woerdeman, M. W. 1937. De tegenwoordige stand van het vraagstuk der embryonale inductie. *Ned. Tijdschr. Geneesk. 81*, 1614.

Woerdeman, M. W. 1938a. Embryonale "inductie" en organisatie. *Vakblad v. Biologen 19*, 97.

Woerdeman, M. W. 1938b. Embryonale Induktion und Organisation. *Biomorphosis 1*, 323.

Woerdeman, M. W. 1950. Over de toepassing van serologische methodes in de experimentele embryologie. *Verslag Kon. Ned. Akad. Wetensch. Amsterdam 59*, 5, 58.

Woerdeman, M. W. 1953. Serological methods in the study of morphogenesis. *Arch. néerl. Zool. Suppl.*

III. IMMUNOGENETICS

BY M. R. IRWIN[1]

THE TITLE of this chapter in a symposium on the general subject of biological specificity is not as formidable as it may appear at first sight. It is, of course, a parallel copy of the term "immunochemistry," which has already attained a place of considerable respectability. The compound word "immunogenetics" indicates that the principles and techniques of genetics and immunology have been used jointly in the prosecution of experiments and in the interpretation of results. In general the title implies that the characters under study are detectable only by immunological techniques, and also that the antigenic characters are gene-determined.

Several review papers on this general topic have been written during the last few years. Therefore needless repetition has been avoided in this chapter and emphasis has been placed on the kind of relationships which may be deduced between the causative genes and their respective antigenic products. The antigenic products of genes which are best known are those of the red blood corpuscles, commonly called red blood cells, and only such antigenic characters will be considered here.

One advantage of studying the specificity existing between the gene and its antigenic product (cellular antigen, antigenic substance, or antigenic character) on red blood cells is that there is as yet no report of a detectable influence of the environment, either external or internal, upon the cellular antigens. It is quite true that the cells of some heterozygotes are somewhat less sensitive in their reactivity with their specific antibodies than are the cells of either homozygote (see Stormont, 1952, for examples in cattle, and Wiener, 1943, for examples in man), but this can hardly be taken as an example of an effect of the environment.

The first cellular antigens within a species which were later shown to be hereditary were the constituents of the A-B-O system of man. The historical account of the development of information about this system, and of the discovery of other "genetic systems" of cellular antigens in man, may be obtained in any one of several review papers or books, for example Wiener (1943) and Race and Sanger (1950). The term "genetic system" is used to describe those cellular antigens whose causative genes are alleles or may be members of a multiple allelic system or are carried on the same chromosome, i.e. linked. (If a separation of the con-

[1] Department of Genetics, University of Wisconsin.

stituents of a genetic system should occur as a result of crossing over of linked genes, it would only be necessary to recognize more than one such system.) The term would also be applied to a cellular antigen which is recognized by its interaction with a specific antibody, but the contrasting cellular antigen or antigens of the genetic system are not thus recognized because the investigators do not have the specific antibodies. That is, in such a genetic system one or more members are not recognized because of the lack of the specific antibodies.

With but two exceptions the cellular antigens known at present in any animal species behave as dominant genetic characters, in contrast to their absence. This does not mean, however, that the "absence" of a detectable antigen in a genetic system should be construed as indicating that this absence is necessarily a recessive character, thereby suggesting that the allelic gene has little or no antigenic effect. As stated above, the presumed absence of a contrasting antigen to one recognized may be simply the inability of the investigator to have at hand, or to find, an antibody specific for it. For example, the co-dominance of the A and B substances of human cells has been universally recognized, but the dominance of each to the O substance has been generally assumed. Recently Boorman et al. (1948) reported evidence for the co-dominance of the O substance with A and B, and the probable relationship of the product of the gene for O to the substances A and B has been indicated by Morgan and Watkins (1948). Partial confirmation of these findings, that a basic constituent (called H) of the A, B, and O substances was responsible for the reactivity of O cells with normal eel (*Anguilla vulgaris*) serum, was obtained by virtue of the reactivities of antibodies in three human sera (Bhende et al. 1952), which in their selective activity for human cells provide additional evidence for the complex antigenicity of the O substance. In brief this example shows very clearly that it is not justifiable to conclude that an antigenic substance is a recessive character unless the criteria from both genetic and serological approaches have been satisfied.

The two possible exceptions to the general principle that the genes affecting cellular antigens produce their respective products without dominance are the Lewis (a$^+$) factor of humans and an "O" substance of sheep. The Lewis antigen was first described by Mourant (1946). It behaves in inheritance as expected of a recessive character (Andresen, 1947; Race et. al., 1949), but since it can be washed off the blood cells (Grubb and Morgan, 1949) it seems much more reasonable to state that this antigen is primarily a secretory product and is only incidentally attached to the blood cells. In this respect it parallels the J substance of cattle cells, which has been shown by Stormont (1949) to be a substance primarily of the plasma which is acquired by the cells from the plasma. The other exception is that reported in sheep by Stormont (1951) in which the substance called O is recognized by its reactivity with a specific antibody from cattle serum, but cells from individuals heterozygous for the gene for the O substance are not reactive. (This is in direct contrast to the reactivity of both A and B substances in the heterozygote for AB in man to react with their specific antibodies. Other parallel examples could be given.) Experimental results as yet unpublished indicate strongly that the O substance of sheep is the antithetical substance to the R of sheep cells described by Ycas (1949), and that both the O and R substances of sheep parallel the J of cattle in being constituents of the plasma which are taken up by the blood cells. Hence the respective genes for the J of cattle, the R and O of sheep, and possibly for the Lewis factor of man, exert their effect on the plasma, and their presence on the blood cells seemingly is more or less accidental. In the light of our present knowledge, then, it appears that there is no exception to the general principle that the genes directly affecting the *cellular* antigens produce their specific products irrespective of the genetic complex in which they operate.

The sequence of events between the genes and the cellular antigens which are produced appears to be relatively direct. That is, if there is a series of reactions between the gene and the end product, the cellular antigen, there are no known instances in which the chain of reactions may be interrupted, by another gene or otherwise. In general, then, an individual will possess a cellular antigen only if either or both parents possess it, implying thereby that the causative gene came from one or both parents.

I. ANTIGENIC SPECIFICITIES PRESUMABLY DUE TO GENIC INTERACTION

There are, however, a few possible exceptions to this general principle of all cellular antigens being more or less the direct products of their causative genes. One such case was reported in chickens by Thomsen (1936), for which evidence was presented that in one family certain chicks possessed an antigenic specificity not present in either of the parents. However, as described elsewhere (Irwin, 1952) the results may be explained as being due to peculiarities of the experimental technique rather than to genic interaction.

A new antigenic specificity not demonstrable on the cells of either parental species has been noted on the cells of certain of the species hybrids which have been obtained in our laboratory. For example, such a specificity has been observed in all species hybrids from the mating of Pearlneck (*Streptopelia chinensis*) and Ring dove (*St. risoria*) as previously reported (Irwin and Cole, 1936a). This new specificity of the cells of the hybrid was demonstrable by virtue of the agglutination of the cells of the hybrid with an antiserum against the cells of the species hybrid, following absorption of the antiserum with the cells of both parental species. The same result was obtained if the cells of the actual parents of some of the hybrids were used in the absorption, thus ruling out the possibility that the transmission of individual differences from either parental species might be the explanation of the new specificity.

Following backcrosses to Ring dove of the species hybrids and selected backcross hybrids, nine antigenic characters were obtained in unit form, i.e. in mating to Ring dove a backcross bird possessing a single antigenic substance peculiar to Pearlneck the offspring were of two kinds—those with and those without the cellular antigen. The specific antigens of Pearlneck thus obtained in unit form were called ("d" indicating dove) d-1, d-2, d-3, d-4, d-5, d-6, d-7, d-11, and d-12 (Irwin, 1939).

One antiserum (from rabbits) against the cells of the species hybrids showed three distinct specificities to the hybrid substance. One specificity was always associated with the d-4 character, another with the d-11. These results lead to the conclusion that, if the respective hybrid substances associated with the d-4 and d-11 characters of Pearlneck are the products of genic interaction, the linkage between the gene or genes producing either the d-4 or d-11 substance and the particular component of the hybrid substance always associated with one or the other of these Pearlneck antigens has not yet been broken. An additional hybrid substance, indistinguishable and therefore presumably identical, was found to be associated with the d-1, d-2, d-3, d-7, and d-12 antigenic substances. However, the offspring from matings to Ring dove of backcross birds carrying one of these specific substances of Pearlneck and the hybrid substance specificity were of four kinds: (a) those with the Pearlneck character and the fraction of the hybrid substance; (b) those with the Pearlneck substance only; (c) those with the hybrid substance only; and (d) those with neither (Irwin and Cumley, 1945). These data were suggestive of linkage of the causative genes for the respective specific Pearlneck substances and a gene or genes which by interaction produced the hybrid specificity, although the possibility can not be entirely

[58]

excluded that a gene or genes on a single chromosome from Pearlneck was responsible, by interaction with a gene or genes from Ring dove, for the hybrid substance associated with the different Pearlneck specific characters. For present purposes the main point to be emphasized is that the cells of some backcross hybrids were reactive with the reagent for the hybrid substance and therefore possessed the hybrid substance specificity without possessing a demonstrable specific Pearlneck character.

The hybrids between the common pigeon (*Columba livia*) and the Ring dove are characterized by a hybrid substance (Irwin and Cole, 1936b). Furthermore the respective reagents for the hybrid substances of the F_1 from Pearlneck and Ring dove and of the F_1 from *livia* and Ring dove (antisera to the cells of the species hybrids absorbed by the cells of the two parental species) are reactive not only with the cells used in immunization but also with the cells of the other species hybrid (Irwin and Cumley, 1945). These results are indicative of an antigenic similarity of the hybrid substances. This antigenic similarity was found to be primarily if not entirely confined to the antigenic specificity associated with the d-11 substance of Pearlneck. (The various results leading to this conclusion are given in Table I.) That is, absorptions of the reagent for the hybrid substance of the Pearlneck-Ring dove hybrids (a) with the cells carrying d-11 would remove all antibodies for such cells as well as for those of the F_1-pigeon Ring dove, or (b) with the cells of the hybrid between *livia* and Ring dove would reduce the titer of the reagent for the cells of backcross birds carrying the d-11 of Pearlneck, but would not affect the reactions for the cells of other black-cross hybrids carrying other fractions of the hybrid substance associated with different specific substances of Pearlneck. Similarly the reagent for the hybrid substance of the cells of the *livia*-Ring dove hybrid has been reactive with the cells of the hybrids between Pearlneck and Ring dove, and also with those of backcross birds containing the d-11 substance of Pearlneck and therefore the hybrid substance associated with it, but not with cells of backcross birds containing any other part of the hybrid substance associated with cellular antigens peculiar to Pearlneck. These results suggest that there are one or more genes in both Pearlneck and *livia* which, by interaction with a gene or genes in Ring dove, produce antigenic products which are related but not identical.

If the antigenic specificity called the hybrid substance on the cells of these two kinds of species hybrids were the result of some kind of interaction of antigens rather than of genes, it would be expected that an antigenic substance related to the d-11 of Pearlneck would be demonstrable in *livia*. However, the cells of the pigeon contain no demonstrable antigen related to the d-11 of Pearlneck (Irwin, 1953). This result does not deny an interaction of antigens, but makes that explanation of the occurrence of the hybrid substances somewhat less plausible.

TABLE I. Tests for similarities and differences of the respective hybrid substances of two kinds of species hybrids of Columbidae.

Antiserum	Absorbed by cells of	Test cells					
		Pearl-neck	Ring dove	Pigeon	$F_1\dfrac{\text{Pearlneck}}{\text{Ring dove}}$	$F_1\dfrac{\text{Pigeon}}{\text{Ring dove}}$	d-11
$F_1\dfrac{\text{Pearlneck}}{\text{Ring dove}}$	{ Pearlneck { Ring dove	0	0	0	++	++	++
$F_1\dfrac{\text{Pearlneck}}{\text{Ring dove}}$	{ Pearlneck { Ring dove { d-11	0	0	0	+	0	0
$F_1\dfrac{\text{Pearlneck}}{\text{Ring dove}}$	{ Pearlneck { Ring dove { $F_1\dfrac{\text{Pigeon}}{\text{Ring dove}}$	0	0	0	+	0	+
$F_1\dfrac{\text{Pigeon}}{\text{Ring dove}}$	{ Pigeon { Ring dove	0	0	0	+	+	+
$F_1\dfrac{\text{Pigeon}}{\text{Ring dove}}$	{ Pigeon { Ring dove { $F_1\dfrac{\text{Pearlneck}}{\text{Ring dove}}$	0	0	0	0	+	0

Symbols: ++, marked agglutination; +, agglutination; O, no agglutination—all at the first dilution of the serum-cell mixture.

Recently a hybrid substance has been demonstrated by Bryan and Miller (1953) to be present on the cells of all backcross birds heterozygous for genes for a specific substance (called C) of *Columba guinea* and a contrasting substance (called C′) of *C. livia* from a mating of *guinea* and *livia*. Backcross birds homozygous for either C (CC) or C′ (C′C′) did not possess the new specificity of the heterozygotes (CC′) —indeed the cells of homozygotes (CC), which in mating to *livia* (C′C′) were the parents of many heterozygotes (CC′), were used in absorption of the antiserum to prepare the reagent which detected the specificity on the cells of the heterozygotes (CC′).

If the new specificity of the cells of the heterozygotes (CC′) were the result of an interaction of the cellular antigens (C and C′) it is reasonable to expect, but it need not necessarily follow, that the new specificity might be formed by virtue of a change in the specificity of either C or C′ or both. However, it has not been possible to demonstrate that the C or C′ substances in the heterozygote differed qualitatively from those on the cells of the respective homozygotes (CC and C′C′). The only difference observed to date is that a smaller quantity of cells from the respective homozygotes than from the heterozygotes is required to absorb the specific antibodies for C or C′, suggesting only a quantitative difference in the amount of the cellular antigen on the surface of the blood cells.

If the new antigenic specificities of the heterozygotes (CC′) are the result of the interaction of genes, one might go one step further and state that they may well be the result of the interaction of alleles. If this be true, the finding of the new antigenic specificity of the heterozygotes (CC′) would be an example of the interaction product of alleles in heterozygotes and might conceivably be an experimental verification of one explanation of heterosis which assumes that heterozygosity per se is a basis for heterosis. As stated above, the present evidence is to the effect that the genes of the heterozygotes CC′ effect both C and C′, both being qualitatively indistinguishable from the antigens in the homozygotes CC and C′C′, plus the new antigenic specificity.

On the other hand, if the new specificity of the heterozygote is due to some kind of an interaction of the antigens rather than to that of the causative genes, it would seem that this may be a secondary rather than a primary interaction product in heterozygotes. The various possibilities will be more completely discussed elsewhere. In this connection it may be pointed out that the antigens of the blood cells may well be of a different class from those of other tissues or fluids of the body. Hence the hybrid substance in extracts of *Drosophila melanogaster* reported by Fox (1949), presumably the product of the interaction of the wild-type alleles of the genes for ruby (*rb*) and for vermilion (*v*), or of genes closely linked to them (Fox and White, 1953) need not parallel such substances of the blood cells in vertebrates.

II. MULTIPLE ANTIGENIC SPECIFICITIES OF GENES

The demonstration of the various "hybrid substances" described above was a part of a study of the antigenic relationships among species which hybridize. Following successive backcrosses of the species hybrids and selected backcross hybrids to one of the parental species, many if not all of the antigenic substances peculiar to one species were obtained in unit form. The number of these species-specific substances varied from less than six in a comparison of two species of Columba (Irwin, Cole, and

Gordon, 1936) to nine or ten in other species of Columbidae (Irwin, 1939; Irwin and Cumley, 1947). If the genes in a species effecting the cellular antigens which are species-specific to one species in contrast to another, as well as those effecting the antigens shared by the two species, should have alleles with contrasting effects, the number of combinations of the cellular antigens to make for individual differences within a species would be very great. Experimental tests of this possibility in both cattle and chickens have given results according to expectation, in that to date the number of known antigenic differences within these species is so great that it should soon be possible to demonstrate complete, or almost complete, individuality of all members of the species. The rapid increase since 1940 in the number of cellular antigens of man points in the same direction (see Race and Sanger, 1950, for a listing of the different antigens of the blood cells of man.)

Forty-odd antigenic specificities have been recognized to date on the blood cells of cattle. Each antigenic specificity is demonstrable by virtue of the reactivity (lysis) of cells containing it in combination with a specific reagent. The first substance recognized was called A, the next B, and so on to Z, the next A′ (with no relationship indicated to A), B′, ..., L′, and Z′. The recognition of each antigenic specificity is of course dependent upon the preciseness of reactivity of the cells with the particular reagent. As outlined by Ferguson (1941) and by Ferguson et al. (1942) the immunological criterion that each reagent reacted with a single substance was that the corpuscles from any individual with cells reactive with a particular reagent would absorb completely the antibodies from that reagent for the reactive cells of all individuals. Furthermore the genetic criterion that only a single specificity was involved was that the data should fit the expectancy by the gene frequency method, as illustrated below.

Type of mating	Number of offspring	
	With A	Lacking A
A × A	217	23
A × —	76	51
— × —	0	41

Each antigenic factor detectable to date has been present on the cells of an individual only if one or both parents possessed it. If each of these 40-odd factors were the product of a single gene, the number of possible combinations would indeed be very great. However, it was noted that there was an association between certain of these factors. For example Ferguson (1941) pointed out that the specificities C and E did not act independently, that C might appear alone but E was detectable only in the presence of C. An allelic series of genes was proposed to explain the observations; one (C_1) gene producing the combined specificities C and E, one (C_2) for C, and another (c) for the absence of both. Additional antigenic factors (R, W, X_1, X_2, and L′) have been identified as belonging to the C system, as are listed in Table II. The various combinations in which these have been observed have been given in another paper (Stormont, Owen, and Irwin, 1951).

The most complex of the ten genetic systems listed in Table II is

TABLE II. Antigenic factors recognized in genetic systems of cattle blood.

A	B	C	F	J	L	S	Z	H′	Z′
A	B	C_1,C_2	F	J	L	S	Z	H′	Z′
H	G	R	V	—	—	U_1,U_2	—	—	—
—	I	W				—			
	K	X_1,X_2							
	O_1,O_2,O_3	L′							
	P	—							
	Q								
	T_1,T_2								
	Y_1,Y_2								
	A′								
	D′								
	E'_1,E'_2,E'_3								
	I′								
	J′								
	K′								
	—								

Symbols: A dash (—) indicates the absence of a recognizable antigenic factor in each genetic system except in the F system.

In the B system, antigenic factors B, G, I, O_1, O_2, O_3, T_1, E'_1, E'_3, and I′ have occurred alone as well as in various combinations.

In the C system, antigenic factors C_1, C_2, W, X_1, X_2, and L′ have occurred alone and also in various combinations.

that called the B system, which contains 21 antigenic factors, including the subtypes (O_1, O_2, O_3; T_1, T_2; Y_1, Y_2; and E'_1, E'_2, and E'_3). Ten (B, G, I, O_1, O_2, O_3, T_1, E'_1, E'_3, and I′) of these 21 factors have appeared alone as well as in various combinations, the other eleven (K, P, Q, T_2, Y_1, Y_2, A′, D′, E'_2, J′, and K′) have been found only in combination with one or more other antigenic factors. If these antigenic specificities (factors) of the B system were products of the same number of loci as there are antigenic factors, the number of combinations possible would be 2^{21} in contrast to the 80 to 100 combinations recognized at present (Stormont, Owen, and Irwin, 1951, and unpublished data).

The association of these antigenic factors may be illustrated by citing a few examples. Thus the factors B and G may appear either alone or together on the cells of individuals. If either appears alone, as B, the offspring of such animals in mating to others not possessing B will have or will not have B in proportions in accordance with the homozygosity or heterozygosity of the parent with B. Individuals possessing both B and G on their cells in mating to those not having either B or G genetically are of two kinds: one kind will produce offspring half of which will possess B, the other half G, as expected if the causative genes are alleles and the parent was a heterozygote; the other kind of BG individuals will produce offspring of which all will contain BG, or half of which will contain both B and G and the other half neither, as expected if the two antigenic factors behaved in inheritance as a unit. That is, B and G may be contrasting antigenic specificities in some individuals and behave as a unit in others.

Another antigenic factor called K has been detected on the cells of more than 1,000 individuals only when both B and G are also present. The respective factors of this combination, BGK, are of course detected by virtue of the reactivity of cells containing them with the reagents specific for each, but the combination has always been transmitted to the progeny as a unit, as follows:

Type of mating	Number of offspring	
	With BGK	Without BGK
BGK × BGK	151	44
BGK × —	185	137
— × —	0	760

These results show clearly that there is a genetic association of the antigenic factors B, G, and K. The causative genes appear to be members of an allelic series: one allele produces B, another G, another BG together, another BGK, and another effects none of these three antigenic factors. The possibility of linkage of the causative genes can not be entirely excluded, but appears much less reasonable in explaining the observed results than the assumption of multiple alleles.

The above examples have shown that two antigenic factors, B and G, may be antithetical characters, or be united in a single combination, with or without K. Certain other antigenic factors of the B system have always behaved as contrasting characters—B and J', G and J', I and K, I and T_1, K and T_1—and not as members of a single combination (Stormont et al., 1951). Each antigenic factor of a contrasting pair, however, may often be found in combination with some other factor or factors, thus illustrating the genetic association of the various antigenic factors of the B system.

As was stated earlier, there are 80 to 100 combinations of the 21 antigenic factors of the B system which are recognized at present. The data given in Table III illustrate how some of the antigenic factors of

TABLE III. The separation of antigenic factors of the B system into contrasting antigenic complexes in the offspring of selected sires.

Blood type of sire	Offspring			
	Antigenic complex	Number	Antigenic complex	Number
$BGIO_1T_2Y_2A'$	$BGIO_1T_2A'$	25	O_1Y_2A'	23
$BO_2A'E'_3J'K'$	$BO_2A'E'_3$	35	$O_3J'K'$	32
$BO_1Y_2D'J'K'$	BO_1Y_2D'	26	$O_3J'K'$	24
$BGKE'_2$	$BGKE'_2$	23	G	27
$BGO_1Y_2E'_1$	BO_1	15	$GY_2E'_1$	23
$BGKY_2E'_1$	$BGKE'_2$	14	$GY_2E'_1$	15
$GY_2E'_1$	$GY_2E'_1$	19	b	13

this system have separated into contrasting combinations in the offspring of the seven sires given. The blood type of the first sire listed in the table was $BGIO_1T_2Y_2A'$ and, in matings to cows with blood types which could be readily distinguished from that of the sire, his offspring were of two kinds in approximately equal proportions: those with $BGIO_1T_2A'$ and those with O_1Y_2A'. (In these progeny B and G were inherited together, but of course with other antigenic factors also being present.) The fifth sire given in the table had the blood type $BGO_1Y_2E'_1$ and his offspring were of the two combinations BO_1 and $GY_2E'_1$. (These progeny illustrate the antithetical relationship of B and G, as described earlier.)

The separation of the progeny into two and only two groups of the antigenic factors of each sire listed in Table III is illustrative of the genetic behavior of all the combinations (or blood groups) of antigenic factors recognized at present. The data which have been published (Stormont et al., 1951) and extensive unpublished data are in complete agreement in supporting the general conclusion that the blood groups of the B system behave as units in inheritance. Hence the conclusion seems reasonable that each blood group of the B system is the product of a single gene, which means that a series of 80 to 100 alleles is required in explanation of this number of groups in the B system. The same reasoning and conclusions undoubtedly apply to the C system, with 20-odd groups recognized in it. Parallel findings leading to the same general conclusions have been made in chickens by Briles et al. (1950).

The demonstrations in Drosophila that multiple allelic series of genes are in reality composed of "pseudoalleles," or closely linked genes with similar effects, should make one suspect that parallel findings may be made for other multiple allelic series, including the B and C systems of cattle blood (for references to these experimental findings, see Lewis, 1952). However, even if two or more loci affecting antigenic factors of cattle blood should be found to be closely linked in either or both the B and C systems, it also appears very reasonable to anticipate that multiple alleles will exist at each locus.

If it be granted that multiple alleles at one or more loci affect the complexes of antigenic factors in the B and C systems of cattle blood, it follows that each gene is responsible for an antigenic product which may contain several antigenic factors or specificities. Presumably also some idea of the complexity of each of the causative genes may be inferred from the complexity of its product. (The principle emphasized as a result of these studies on cattle blood is the same as that given by Wiener and Wexler (1952) that the antigenic product of a single gene may have several antigenic specificities.)

III. CELLULAR ANTIGENS IN DEVELOPMENT

Until recently it would have been reasonable to believe that all antigens of the blood cells were present on the cells of the embryos. For example the A and B substances of man have been found as early as the thirty-seventh day in the developing fetus (see references in Wiener, 1943); and Keeler and Castle (1933) demonstrated that a contrasting pair of cellular antigens in the rabbit were detectable at the 4 mm. stage. Exceptions to this general rule have been observed, in one instance even within one genetic system. Thus, Briles et al. (1948) have described four antigenic substances within a genetic system of antigens in chickens, one of which could be detected on the cells of the embryo early in the fourth day of incubation whereas the other three were demonstrable only after the chicks had hatched. These and observations by other workers on differences in the time of development of the cellular antigens suggest a parallel difference in the activity of the causative genes—presumably in different blood forming tissues—during embryonic development.

An unusual kind of cellular antigen has been described in cattle for the substance called J. In brief this antigenic substance of the cells appears to be primarily a normal constituent of the plasma of individuals possessing the causative gene and is often but not always taken up by the cells. Stormont (1949) demonstrated that cells from an animal not possessing J could "acquire" them either *in vivo* or *in vitro*. And Stone (1953) showed that the J substance might be found only in the plasma, not on the cells, of some individuals, and in the plasma and on the cells of others. The J substance was always present at birth in the plasma of each calf carrying the causative gene whether or not it appeared on the cells at a later date. Inquiries into the reasons why (a) the cells of the newborn do not possess the J substance although it is present in the serum, and (b) why the cells of some individuals never acquire demonstrable J substance although it is present in the serum, are indicated. A substance with many attributes paralleling those of the J of cattle has been described in sheep (Ycas, 1949). And as stated earlier the Lewis factor of human cells seemingly is another example of the antigenic effect of a gene on the plasma which is recognizable because it is (loosely) affixed on the cells.

IV. RECIPROCAL TRANSPLANTATION OF FETAL TISSUE IN CATTLE TWINS

In man, identical twins always have the same blood type in contrast to fraternal twins, which have no more similar blood types than sibs. It was quite unexpected then to find twins with identical blood types (Owen, 1945) resulting from a case of superfecundation in which a Guernsey cow had given birth to twins with different markings. One twin had the typical Guernsey color, the other had a white face, and the herd records showed that the cow had first been bred by a Guernsey bull and later by a Hereford bull. The blood types of these five animals are as follows:

Guernsey sire:	A	H I	S X_2 Z	E' H'
Hereford sire:	A C E		R W	I'
Guernsey dam:	A C E G	Q	W Z D' E' H'	Z'
Guernsey twin—male	A C E G	Q R S W X_2 Z D' E' H' I' Z'		
Hereford twin—female	A C E G	Q R S W X_2 Z E' E' H' I' Z'		

Not only were the blood types of the twins identical but also the Guernsey twin contained antigenic factors R and I', the genes for which could have been transmitted only by the Hereford sire, and the Hereford-marked twin possessed factors S and X_2, the genes for which could have come only from the Guernsey sire. Subsequent tests on other twins in cattle by Owen (1945) and others have shown that about 90 per cent have identical blood types except for the antigenic substance J (Stormont, 1949).

The most rational explanation of these findings, as proposed by Owen (1945), is that the union of the circulatory systems in about 90 per cent of bovine twins (Lillie, 1916) provides a bridge for the reciprocal migration of the so-called primordial blood cells and these become established in the hematopoietic tissues of the co-twin. Thereafter they produce blood cells and so each twin throughout life has a mosaic of blood cells—those from its own blood forming tissues and those from the tissues of the co-twin. A differentiation of the two kinds of cells can be made in the laboratory, thus demonstrating that there are two kinds of blood cells, usually in unequal proportions, in each such twin. Very peculiarly, each twin has been found to have the same proportion of each of the two kinds of blood, so in the majority of twins one of the pair has more of its co-twin's blood than it has of its own.

Another unusual finding in cattle twins has been made by English workers in attempting to use the techniques of skin grafting in the differentiation of identical and non-identical twins. From the experience in other species of animals it would be anticipated that identical twins in cattle would accept grafts of each other's skin but non-identical twins would not. The reports by these workers (Anderson et al., 1951; Billingham et al., 1952) that dizygotic twins in cattle will usually accept each other's skin graft, whereas sibs will not, suggest that the reciprocal migration of primordial blood cells resulted in more of an effect than just the admixture of bloods. One might hypothesize, but not in explanation of this phenomenon, that the migration of embryonal cells need not be restricted to the primordial blood cells and that any primordial cell which could migrate might be expected to establish itself in the co-twin.

A few examples have been discussed of the biological specificity of gene products, possibly primary gene products, which are recognized by immunological methods. These gene products may be studied further to gain more information about the basis of their biological specificity such as the multiple antigenic specificities of the products of single genes. Furthermore it seems probable that they may also serve as tester substances for various problems of development.

BIBLIOGRAPHY

Anderson, D., R. E. Billingham, G. H. Lampkin, and P. B. Medawar. 1951. The use of skin grafting to distinguish between monozygotic and dizygotic twins in cattle. *Heredity 5,* 379-397.

Andresen, P. H. 1947. Blood group with characteristic phenotypical aspects. *Acta Path. Microbiol. Scand. 24,* 616-618.

Bhende, Y. M., C. K. Deshpande, H. M. Bhatin, R. Sanger, R. R. Race, W. T. J. Morgan, and W. M. Watkins. 1952. A new blood-group character related to the ABO system. *Lancet* (May 3), 903-904.

Billingham, R. E., G. H. Lampkin, P. B. Medawar, and H. LL. Williams. 1952. Tolerance to homografts, twin diagnosis and the freemartin condition in cattle. *Heredity 6,* 201-212.

Boorman, K. E., B. E. Dodd, and B. E. Gilbey. 1948. A serum which demonstrates the co-dominance of the blood-group gene O with A and B. *Ann. Eugenics 14,* 201-208.

Briles, W. E., W. H. McGibbon, and M. R. Irwin. 1948. Studies of the time of development of cellular antigens in the chicken (abstract). *Genetics 33, 97.*

Briles, W. E., W. H. McGibbon, and M. R. Irwin. 1950. On multiple alleles effecting cellular antigens in the chicken. *Genetics 35, 633-652.*

Bryan, C. R. 1953. Genetic studies of cellular antigens in Columbidae. Doctor's thesis, University of Wisconsin.

Bryan, C. R., and W. J. Miller. 1953. Interaction between alleles affecting cellular antigens following a species cross in Columbidae. *Proc. Nat. Acad. Sci. 39,* 412-416.

Ferguson, L. C. 1941. Heritable antigens in the erythrocytes of cattle. *J. Immun. 40,* 213-242.

Ferguson, L. C., C. Stormont, and M. R. Irwin. 1942. On additional antigens in the erythrocytes of cattle. *J. Immun. 44,* 147-164.

Fox, A. S. 1949. Immunogenetic studies of *Drosophila melanogaster.* II. Interaction between the rb and v loci in production of antigens. *Genetics 34,* 647-664.

Fox, A. S., and T. B. White. 1953. Immunogenetic studies of *Drosophila melanogaster.* III. Further evidence of genic interaction in the determination of antigenic specificity. *Genetics 38,* 152-167.

Grubb, R., and W. T. J. Morgan. 1949. The "Lewis" blood group characters of erythrocytes and body fluids. *Brit. J. Exp. Path. 30,* 198-208.

Irwin, M. R. 1939. A genetic analysis of species differences in Columbidae. *Genetics 24,* 709-721.

Irwin, M. R. 1952. Specificity of gene effects, in *Heterosis,* ed. by J. W. Gowen. Iowa State College Press, Ames.

Irwin, M. R. 1953. Evolutionary patterns of antigenic substances of the blood corpuscles in Columbidae. *Evolution 7,* 31-50.

Irwin, M. R., and L. J. Cole. 1936a. Immunogenetic studies of species and species hybrids in doves, and the separation of species-specific substances in the backcross. *J. Exp. Zool. 73,* 85-108.

Irwin, M. R., and L. J. Cole. 1936b. Immunogenetic studies of species and of species hybrids from the cross of *Columba livia* and *Streptopelia risoria. J. Exp. Zool. 73,* 309-318.

Irwin, M. R., L. J. Cole, and C. O. Gordon. 1936. Immunogenetic studies of species and of species hybrids in pigeons, and the separation of species-specific characters in backcross generations. *J. Exp. Zool. 73,* 285-308.

Irwin, M. R., and R. W. Cumley. 1945. Suggestive evidence for duplicate genes in a species hybrid in doves. *Genetics 30,* 363-375.

Irwin, M. R., and R. W. Cumley. 1947. A second analysis of antigenic differences between species in Columbidae. *Genetics 32,* 178-184.

Keeler, C. E., and W. E. Castle. 1933. A further study of blood groups of the rabbit. *Proc. Nat. Acad. Sci. 19,* 403-411.

Lewis, E. B. 1952. The pseudoallelism of white and apricot in *Drosophila melanogaster. Proc. Nat. Acad. Sci. 38,* 953-961.

Lillie, F. R. 1916. The theory of the free-martin. *Science 43,* 611-613.

Morgan, W. T. J., and W. M. Watkins. 1948. The detection of a product of the blood group O gene and the relationship of the so-called O substance to the agglutinogens A and B. *Brit. J. Exp. Path. 29,* 159-173.

Mourant, A. E. 1946. A "new" human blood group antigen of frequent occurrence. *Nature 158,* 237-238.

Owen, R. D. 1945. Immunogenetic consequence of vascular anastomoses between bovine twins. *Science 102,* 400-401.

Race, R. R., and Ruth Sanger. 1950. *Blood Groups in Man.* Blackwell Scientific Publications, Ltd., Oxford, and Charles C. Thomas, Springfield, Ill.

Race, R. R., R. Sanger, S. O. Lawler, and D. Bertinshaw. 1949. The *Lewis* blood groups of 79 families. *Brit. J. Exp. Path. 30,* 73-83.

Stone, W. H. 1953. Immunogenetic and immunochemical studies of an antigenic character in cattle blood. Doctor's thesis, University of Wisconsin.

Stormont, Clyde. 1949. Acquisition of the J substance by the bovine erythrocyte. *Proc. Nat. Acad. Sci. 35,* 232-237.

Stormont, Clyde. 1951. An example of a recessive blood group in sheep (abstract). *Genetics 36,* 577-578.

Stormont, Clyde. 1952. The F-V and Z systems of bovine blood groups. *Genetics 37,* 39-48.

Stormont, C., R. D. Owen, and M. R. Irwin. 1951. The B and C systems of bovine blood groups. *Genetics 36,* 134-161.

Thomsen, O. 1936. Untersuchungen über erbliche Blutgruppenantigene bei Hühnern, II. *Hereditas 22,* 129-144.

Wiener, A. S. 1943. *Blood Groups and Transfusion*. Charles C. Thomas, Springfield, Ill.

Wiener, A. S., and I. B. Wexler. 1952. The mosaic structure of red blood cell agglutinogens. *Bact. Rev. 16*, 69-87.

Ycas, M. K. W. 1949. Studies of development of a normal antibody and of cellular antigens in the blood of sheep. *J. Immun. 61*, 327-347.

IV. ENZYME DEVELOPMENT AS ONTOGENY OF SPECIFIC PROTEINS

BY S. C. SHEN[1]

THE major premise in chemical embryology is that morphogenetic changes are basically chemical changes and are therefore amenable to chemical analysis and interpretation. The intricate events in morphogenesis may possibly be integrated as a chain of chemical reactions leading to the formation of end products that characterize adult structures and functions. Since proteins are the principal elements of organic forms, it is further assumed that the basic chemical changes in morphogenesis involve the production and transformation of protein molecules and their aggregates. Thus the ontogeny of specific structures and functions may be described as the ontogeny of specific proteins.

Conceptually the postulation appears to be entirely plausible and even obvious. The experimental substantiation of these assumptions is, however, a highly complex and devious task. The immediate questions to be raised are: What are these specific proteins? How do they change in the course of development? How are they related to morphological and functional differentiations? Finally how are they formed?

I. ENZYMES AS SPECIFIC PROTEINS

Although the chemistry of developing embryos has been fairly extensively investigated, little is known of the nature of their protein contents. This is of course not surprising in view of our severely limited knowledge of protein chemistry. In spite of the impressive progress in recent years in the physics and chemistry of proteins, the precise constitution and structure of a protein molecule remains largely obscure. The qualitative and quantitative analyses of individual proteins present problems of considerable technical difficulty. In some instances a protein can be characterized by its dynamic properties rather than by its molecular construction. Recent advances in the study of protein specificity by its immunological property are of special interest to biologists in general and embryologists in particular. The immunobiological approach to developmental problems is discussed in Chapters I and II of this symposium. The present paper

[1] Osborn Zoological Laboratory, Yale University. The original investigations reported by the author in this chapter have been supported in part by a grant from the American Cancer Society through the Committee on Growth of the National Research Council.

attempts to consider the possible significance of the enzymological aspect of protein specificity in relation to biological differentiation.

The significance of the appearance and development of enzymatically active and specific proteins during morphogenesis is at least twofold. In the first place specific enzymes are most probably involved in the synthesis and transformation of specific proteins. From this point of view the study of developmental enzymology contributes directly to a causal analysis of proteinogenesis as a chemical basis of morphogenesis. Unfortunately the serious limitation of our present day knowledge of protein synthesis makes it premature to contemplate the role of enzymes as the regulatory factors in ontogenetic changes. For the present therefore, the developmental significance of enzymes is to be considered on an extremely limited basis. The enzymatic properties are employed primarily as descriptive indices of certain tissue proteins that are being formed in the course of embryonic development. The qualitative and quantitative aspects of enzyme analysis in a developing system are therefore descriptions of the "facts" rather than "factors" of the morphogenetic process. The study of developmental enzymology, in this restricted sense, is essentially descriptive in nature and has in itself no explanatory value in the chemical analysis of morphogenesis. Nevertheless in chemical embryology, as in classical embryology, accurate and adequate descriptions of a phenomenon are the indispensable groundwork without which experimental and causal analyses of morphogenetic processes are neither possible nor valid.

II. ENZYMES AS FUNCTIONAL AND STRUCTURAL PROTEINS

A detailed account of certain chemical changes that occur during the course of embryonic development is not necessarily a chemical description of differentiation. The study of any chemical aspect of morphogenesis is a relevant and valid approach to embryological problems only if it is undertaken with a view to ascertain a possible common ground for chemical and biological characteristics and manifestations. The rational importance of this point has often been overlooked in many a conscientious investigation in which the chemical analysis of developing embryos is undertaken as a chemical rather than a biological problem. In the study of enzyme development as an embryological problem it is important to consider first the possibility that enzymes are functionally and structurally specific constituents of tissue proteins and may therefore be a significant basis of physiological and morphological specificity.

That specific enzymes are often associated with specific functions is a well-known fact in physiology. In such instances it would seem reasonable to expect the appearance and accumulation of a certain enzyme or enzyme complex to correspond closely to the development or elaboration of a specific function. This is of course but a supposition that is not to be taken for granted but requires experimental verification. The establishment of a definite and precise relationship between enzyme development and physiological differentiation not only is of basic importance in chemical embryology but also contributes directly to the study of enzymology.

The possible role of an enzyme as a structural protein, on the other hand, is at present far less obvious. Until recently this has been inconceivable in view of the classical concept of an enzyme as a protein of negligible physical dimensions. However, the discovery of the adenosinetriphosphatase activity of myosin, a major component of muscle proteins, led Engelhardt to believe that myosin is the enzyme adenosinetriphosphatase itself. It is a staggering idea that an enzyme may constitute a substantial proportion of the total proteins of a tissue. Although there have since been important revisions in the biochemistry of muscle proteins and enzymes, the possibility remains that muscle adenosinetriphosphatase as an enzyme is also one of the principal structural proteins in a contractile tissue.

In addition to the unconventional concept of the size and structure of a particular enzyme there is the intriguing postulation first put forth by Engelhardt that all tissue proteins are catalytically active and specific, though the nature of the specific activities is yet to be discovered. This may represent an extreme view but is not altogether inconceivable. New enzymes are constantly being discovered and new reactions become known that are catalyzed by yet unidentified enzymes. While all proteins are not necessarily enzymes, enzymatically active and specific proteins will undoubtedly become increasingly recognized as extensive and important constituents of living protoplasm. The developmental significance of enzymes as structural proteins may hence become increasingly apparent.

III. METHODOLOGICAL CONSIDERATIONS IN THE ANALYSIS OF ENZYME DEVELOPMENT

The principal feature of biological differentiation is the irreversible emergence of a new and specific pattern of organization with respect to either the embryo as a whole, a specialized organ, tissue, or cell. Chemically, at a given level of differentiation, a new pattern could theoretically arise from one or often both of two changes: (1) quantitative changes in one or more of the constituents resulting in an altered over-all pattern, and (2) spatial changes in the orientation and distribution within the system of its components with or without concomitant quantitative modification of the latter. In treating enzymes as the principal chemical constituents of a biological entity, the methodology in the analysis of chemical differentiation consists essentially in quantitative assays of the enzyme contents with respect to (1) their courses of development, and (2) their spatial distribution within the differentiating system.

In the quantitative estimation of enzymes the term "content" is here used specifically and distinct from "activity." Unlike smaller and relatively simple molecules, enzymes can not at present be measured quantitatively in molar concentrations or even in gravimetric units. The relative amount of an enzyme is only indirectly determined and expressed by the intensity of its specific activity. Since the activity of an enzyme is a function not only of the amount of enzyme but also of many other factors, the quantitative assay of enzyme content is valid only when made under such optimal conditions that the only limiting factor of the reaction is the enzyme concentration. Furthermore, inasmuch as the specific activity of an enzyme can usually be expressed only in arbitrary units, its quantitative significance depends strictly on an explicit definition of the activity unit in which it is measured. These methodological considerations, while rather obvious, are often neglected. Many of the apparent discrepancies and much of the confusion in the experimental findings by different authors could probably have been avoided or materially reduced through proper standardization of the conditions of enzyme measurement and in the terms of expressing enzyme activity.

The quantitative variation of an enzyme in a given system during development includes the possible *de novo* or epigenetic synthesis of an enzyme not previously present and conversely the total disappearance of an existing enzyme. However, the inherent limitation of the present techniques of enzyme assay does not permit a negative result to be interpreted as the total absence of an enzyme. The first detection of an enzyme in a developing system does not therefore necessarily indicate the first appearance of that enzyme. Thus far there is no convincing way of distinguishing the epigenetic or preformed origin of an enzyme. In any case the persistent controversy between epigenesis and preformation has no immediate bearing on the present analysis, which is concerned primarily with the relative change of a status rather than its origin. It may be mentioned in this connection that the notion of a morphologically and functionally undifferentiated embryo as enzymatically void has long been shown to be totally untenable. With improved techniques of enzyme analysis, an increasing variety of enzymes have been demonstrated to be present, although generally at a relatively low level, since the earliest stages of embryonic development. It seems quite probable that the essential feature of chemical differentiation is not so much the total absence or *de novo* appearance of a specific enzyme as the change in relative concentration of the component enzymes through differential rates of accumulation. The process is essentially analogous to morphological heterauxesis as the origin of a new pattern of structural and functional organization.

Perhaps no enzyme is uniformly distributed in an embryo at any stage of its development. The progressive divergence in regional enzyme contents, quantitatively as well as qualitatively, as a measure of chemical heterogeneity is undoubtedly an essential characteristic in chemical differentiation. The specificity in the pattern of enzyme distribution may be a significant expression of the specificity of structural and functional organization at a given level of differentiation. Quantitative assays of enzyme content in whole embryos alone, as have often been done, obviously would yield only limited information on the developmental significance of an enzyme in any particular phase of the morphogenetic process. Similarly the enzyme content of an entire organ is not immediately relevant to the chemical process of organogenesis. Continued progress in the methodology of enzyme localization at an increasingly precise level of biological organization is indispensable to future investigations of chemical differentiation.

Finally, the direct correlation of the chemical with the biological aspect of ontogeny remains a crucial task in chemical embryology that has yet to be explored. Correlations based solely on an apparent and generally crude parallelism between chemical and biological events in the course of normal development may often be fortuitous in character. Casual observations must therefore be supplemented by experimental modifications of the normal developmental process in order to ascertain the relative dissociability of the two aspects of differentiation. Furthermore significant and precise correlations can only be expressed in quantitative terms. The general lack of quantitation of biological differentiation is a major stumbling block in the present attempt.

The complexity of some of these problems of technique as well as of interpretation are illustrated and discussed in the following examples of studies in enzyme development in relation to the differentiation of

[77]

two of the most specialized structures, namely the muscle and the central nervous system. The purpose in citing these studies is patently to show the gross inadequacies of our present attempts and the need for further investigation.

IV. DEVELOPMENT OF ADENOSINETRIPHOSPHATASE

The presence of adenosinetriphosphatase in muscle and its essential role in muscular contraction have long been known. The far-reaching significance of the recent discovery (Engelhardt and Ljubimova, 1939; Szent-Györgyi, 1951) of the apparent identity of the enzyme with the contractile protein myosin has already been discussed. Developmentally, muscle adenosinetriphosphatase thus appears to be a unique enzyme in that as a specific protein it is not only functionally but also structurally significant in the differentiation of contractile tissues.

Moog (1947) first investigated adenosinetriphosphatase activity in developing chick muscle. It was found that the enzyme activity per unit weight of muscle homogenate increased significantly from a relatively low level from the twelfth day of incubation, reached a peak at hatching, and then declined slightly to the adult level. No apparent correlation could be found between histological differentiation or the onset of muscular contraction and enzyme development. It was interpreted that the height of enzyme activity at hatching coincided with the first functional demands on the contractile tissue. Herrmann and Nicholas (1948) made a similar study in developing rat muscles. The specific enzyme activity was also assayed on muscle homogenates. It was found to remain at a low level up to the fifth day of gestation and then to rise sharply. The increase continues throughout the intra-uterine development and for more than two weeks after birth. The onset of the sharp rise in adenosinetriphosphatase activity appears to coincide with that of muscular function.

A large discrepancy was soon discovered, however, between the enzyme content and the amount of contractile protein recoverable from muscle homogenates. Throughout development, particularly during the early phase of muscle differentiation, the rate of enzyme accumulation far exceeds that of actomyosin, with which the enzyme is presumably associated. To investigate this apparent anomaly, Herrmann, Nicholas, and Vosgian (1949) studied the distribution of adenosinetriphosphatase activity in various fractions of the muscle homogenate. The enzyme was found in all fractions prepared. In fact only a relatively minor proportion of the total activity is due to the myosin fraction at all stages of development. Although it is well known that adenosinetriphosphatase activities are found in a wide variety of tissues other than skeletal muscle, the presence in muscle of a nonmyosin adenosinetriphosphatase has been demonstrated by Kielley and Meyerhof (1948). It is enzymatically distinct from the myosin adenosinetriphosphatase by the following characteristics: (1) it is Mg^{++} activated and depressed by Ca^{++}; (2) it has a pH optimum at 6.8; and (3) it is associated with the particulate matters of muscle extract.

In view of these findings the development of adenosinetriphosphatase or adenosinetriphosphatases in rat muscle has been reinvestigated (de-Villafranca, 1953). By appropriate methods of fractionation the muscle homogenate was separated into relatively distinct fractions, principally actomyosin and nonmyosin particulates of various sedimentation characteristics. Throughout development, as in the adult, highest adenosinetriphosphatase activities are found in the actomyosin and in the "small granules," probably microsomes, fractions of the muscle homogenate. The actomyosin adenosinetriphosphatase is consistently activated by Ca^{++}, only slightly affected by Mg^{++}, and exhibits a characteristic alkaline pH optimum in the presence of appropriate buffers (Bailey, 1942). In contrast to this the "particulate" adenosinetriphosphatase is characterized by Mg^{++} activation, Ca^{++} depression, and a pH optimum at 6.8. Under their respective optimal conditions of pH and ion activation the Mg adenosinetriphosphatase is several times higher in specific enzyme activity than the Ca or myosin adenosinetriphosphatase. The appearance and development of the two adenosinetriphosphatases in muscle is shown in Fig. 1. The activity of both enzymes remains at a low level up to the fifteenth day of gestation, at which point both start to increase rapidly and steadily. However, they differ somewhat in their courses of development after their simultaneous initial rise. The particulate adenosinetriphosphatase shows a much steeper rate of increase at the earlier stages of development, reaches a peak at approximately 40 days after birth, and maintains a constant level thereafter. Between birth and the fortieth day postpartum its specific activity has increased nearly tenfold as compared with a fourfold increase in the myosin adenosinetriphosphatase during the same period of development. The latter, however, continues to increase without reaching a steady level for several months after birth.

At least two points of interest have emerged from these findings. In the first place it is clear that in developing, as in adult muscle, there are present at least two adenosinetriphosphatases that are distinct in their enzymatic characteristics as well as in their spatial relationship with the

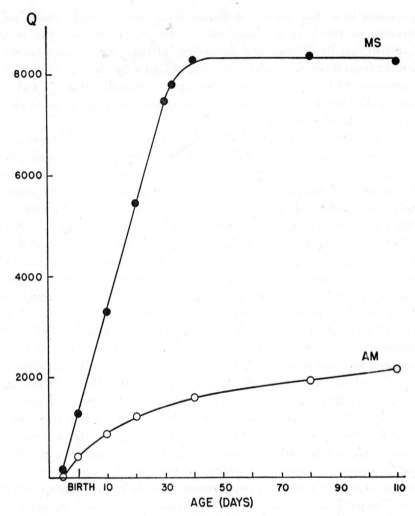

Fig. 1. Development of adenosinetriphosphatase activity in the actomyosin (AM) and in the "microsome" (MS) fractions of rat muscle. The specific enzyme activity Q is expressed as micrograms of P liberated in 15 minutes per mg. protein nitrogen. Myosin adenosinetriphosphatase is maximally activated by 0.01 M $CaCl_2$ and particulate adenosinetriphosphatase by 0.005 M $MgCl_2$.

contractile elements. While it seems quite certain that myosin adenosine-triphosphatase is directly and specifically involved in the contractile proc-ess, the functional significance of the particulate adenosinetriphosphatase in the general metabolism of muscle is not yet known. Presumably it is involved in an energy generating reaction necessary for activities other than muscular contraction. Whether or not it plays an important role in the synthesis of specific proteins such as actomyosin during development is a matter for idle speculation at the present. The second point of interest is that the rate of enzyme accumulation does not bear a constant relation with that of any isolated relatively homogeneous fraction of the muscle proteins. In the case of myosin adenosinetriphosphatase, the increase in its specific activity in the actomyosin fraction may conceivably be the reflection of a progressive change in the myosin-actin ratio. The relative constancy of such a ratio in adult muscle does not necessarily extend to embryonic muscles. It is of interest to note that actomyosin preparations from embryonic or newborn rats differ markedly from that of adult rats in their physical characteristics (Ivanov and Kassavina, 1948). The former show no viscosity drop upon the addition of ATP; they do not readily form actomyosin threads, and when they do the resulting threads do not respond to treatment of ATP. The difference in these properties may well be the result of a difference in the actin-myosin makeup of the contractile protein. When it becomes possible to prepare chemically pure myosin from developing muscles, it would then be possible to test crucially the dissociability of the enzyme adenosinetriphosphatase from myosin. Recent evidence tends to show that only part of the myosin molecule has the adenosinetriphosphatase activity (Mihályi and Szent-Györgyi, 1953). Therefore it appears at least theoretically possible that the development of adenosinetriphosphatase activity in muscle may involve a genuine intramolecular differentiation of the protein myosin.

The nature of the particulate matter with which the Mg adenosinetriphosphatase is apparently associated is not yet completely known. The enzyme is generally believed to be part of the mitochondrial complex but becomes active *in vitro* only when the mitochondria are structurally disrupted (Sacktor, 1953). The accumulation of Mg adenosinetriphosphatase in a given species of cell particulate may represent differentiation within a cytological constituent. The significance of chemical differentiation on a cytological level will be discussed more fully in the following section.

No positive correlation can as yet be deduced between the development of either of the two adenosinetriphosphatases and muscular differentiation. The fortuitous character of the apparent agreement between the initial sharp rise in enzyme content and the beginning of active histological and functional differentiation can not be ruled out at the present.

V. DEVELOPMENT OF RESPIRATORY ENZYMES IN MUSCLE
MITOCHONDRIA

One of the most significant advances in cytochemistry in recent years
has been the enzymology of cytological elements, notably mitochondria.
Not only are certain enzymes in a cell found exclusively associated with
the mitochondrial particles but also the coexistence of these enzymes in
a single cytological unit strongly indicates a high degree of structural
integration within the unit. The fundamental importance of this concept is that for the first time an enzyme complex, such as the cyclophorase
system of Green, may now be viewed not as a mere conglomeration of
crystalline proteins but as a tangible cytological entity. The bearing of
this on the chemistry of biological differentiation is only too obvious.

Among the various oxidative enzymes found in mitochondria, cytochrome oxidase and succinic dehydrogenase are the first ones to be
convincingly demonstrated. By appropriate methods of fractionation,
virtually complete recovery of these enzymes can be obtained from the
mitochondrial fraction. The development of these enzymes in rat muscle mitochondria has been studied (Shen, 1949). In the preliminary work
particulates were isolated from saline suspensions. Identical results have
since been obtained by using isotonic sucrose solution as the suspension
medium. These results are shown in Fig. 2. From the fifteenth day of
gestation—the beginning of an active phase of muscular differentiation—
up to the second week after birth there is a rapid increase of these two
enzymes in the mitochondrial fraction. During that period the rate of
increase is exponential and constant throughout. The identical rate of
increase of the two enzymes indicates a closely integrated enzyme system in function as well as in development.

The validity of the interpretation of these results as a chemical differentiation of muscle mitochondria depends to a large extent on experimental evidence for the homogeneity of the mitochondrial preparations.
In the present experiment the apparent homogeneity of the mitochondrial
fractions as prepared from homogenates of muscle at various stages of
differentiation are indicated by the following facts: (1) the absence of
DNA in all such preparations rules out any possible contamination by
nuclear materials which are particularly abundant in embryonic muscles; (2) the optimal sedimentation rate in the differential centrifugation to yield mitochondrial preparations of maximum specific enzyme
activity has been determined and used in each muscle homogenate, and
it was found to be identical in all cases; (3) the mitochondrial fractions

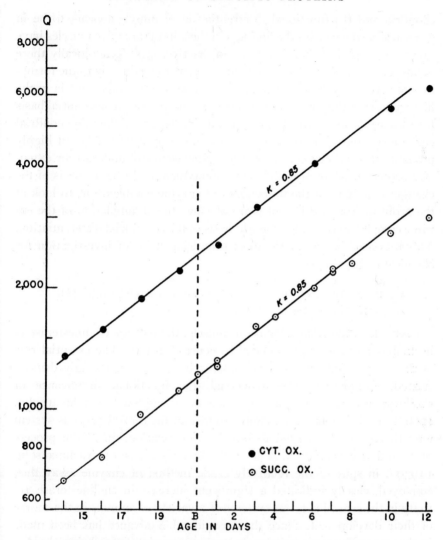

Fig. 2. Development of respiratory enzymes, cytochrome oxidase (CYT.OX.), and succinoxidase (SUCC.OX.) in the mitochondrial fraction of developing rat muscle. B, birth; k calculated according to the equation $X = ae^{kt}$.

were subjected to repeated washings until the specific enzyme activities reached a maximum and remained constant with further washings; (4) direct examination of the preparation under phase contrast microscope revealed the absence of visible contamination by other cytological components or debris.

The apparent parallelism between the enzyme accumulation in mito-

S. C. SHEN

chondria and the functional differentiation of muscle appears to be in general accordance with the findings in the adult that active muscles have greater cyclophorase activity than do inactive ones. Cytologically there is suggestive evidence that mitochondria contribute directly to the formation of myofibrils. Knowledge of the precise role of mitochondria in the histogenesis of the contractile elements would provide a substantial basis for the elucidation of the developmental significance of the mitochondrial enzymes in muscle differentiation. It is not only conceivable but highly probable that mitochondria as cytological elements undergo an extensive process of structural differentiation which might be the basis of the change in activity of the mitochondrial enzyme complex. Or, to look at the problem from a different point of view, the accumulation of the enzyme complex may be the chemical basis of mitochondrial differentiation. This may well be one of the most promising fields of investigation for chemical embryologists.

VI. DEVELOPMENT OF CHOLINESTERASE IN THE CENTRAL
NERVOUS SYSTEM

Like adenosinetriphosphatase in muscle, the enzyme cholinesterase is intimately and specifically related to neural function. The essential role of this enzyme in synaptic transmission has been convincingly demonstrated. Although still controversial, the functional significance of cholinesterase in nerve conduction has been suggested (Nachmansohn, 1950). The development of cholinesterase in the central nervous system may therefore be expected to bear a close relationship to the process of neural differentiation. The early investigations of Nachmansohn (1939), in spite of the relatively crude method of enzyme assay then employed, clearly indicated a significant increase in cholinesterase activity in the central nervous systems of a variety of animals in the course of their development. Since then considerable advance has been made in the enzymology of cholinesterase, the most significant being the distinction of two types, the specific and the nonspecific. It is the specific, or acetylcholinesterase, that is primarily found in the nervous system and functionally associated with it.

With the application of the specific and sensitive method of enzyme analysis, the development of cholinesterase in the central nervous system has been reinvestigated in a number of animals, notably the amphibians. From the classical work of Coghill (1929) it is apparent that in the amphibian the functional differentiation of various parts of the central nervous system occurs at different stages of embryonic de-

[84]

velopment. Based on the motor behavior of the developing animal, it appears that neural differentiation begins at the spinal cord and proceeds sequentially cephalad to medulla, mesencephalon, and finally the hemispheres. If a specific correlation exists between cholinesterase develop-

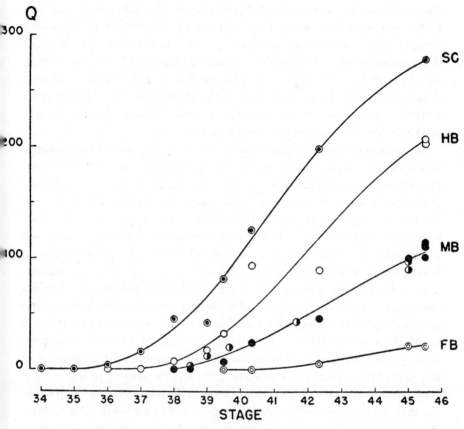

Fig. 3. Development of cholinesterase in various parts of the central nervous system of *Amblystoma punctatum*. FB, forebrain; MB, midbrain; HB, hindbrain; SC, spinal cord. Cholinesterase activity Q expressed as millimicroliters CO_2 per microgram tissue nitrogen per hour. Abscissa in Harrison's developmental stages.

ment and neural differentiation, a similar caudo-cephalad sequence in the enzyme accumulation may be expected. The experimental results (Boell and Shen, 1950) as shown in Fig. 3 are clearly in agreement with this expectation. In the developmental stages, when no sign of neurogenic response is observable, no detectable amount of cholinesterase can be found in any part of the central nervous system. The first measurable amount of the enzyme is found in the spinal cord at a stage when the animal first becomes capable of responding to neural stimulations. Significant enzyme activity begins to appear in the hindbrain when the animal first exhibits spontaneous swinging movement. Well-coordinated and sustained swimming becomes apparent at about stage 40 when cholinesterase begins to make its appearance in the midbrain. This parallelism is particularly interesting in view of the finding of Detweiler (1948) that experimental interference of the midbrain at this stage of development produced the most pronounced effect on the swimming ability of the animal. It suggests that the midbrain at that time assumes a dominant role in the coordination of the motor activities. The enzyme activity appears last in the hemispheres and continues to remain at a relatively low level throughout development.

In sharp contrast to the differential and sequential pattern of accumulation of the enzyme cholinesterase in the parts of the central nervous system, the respiratory activity and the respiratory enzymes, cytochrome and succinic oxidases, are found to be practically identical in all parts of the brain. Furthermore during the critical period of neural differentiation and maturation there is virtually no change in the respiratory enzyme activities. The course of the development of cholinesterase in homogenates of the whole brain is shown in Fig. 4. The first appearance of the enzyme activity is at stage 36 when the animal first exhibits neurogenic responses. In these same homogenates the succinate oxidase activities were also measured (see Fig. 5). It is seen that except for a relatively brief initial rise in enzyme activity in early development, during which there is no discernible sign of functional differentiation in any part of the central nervous system, the specific activity of the respiratory enzyme remains at a constant level. The relatively low activity at the early stages is probably due to the inclusion of a considerable amount of enzymatically inactive yolk material in the primitive brain structure. The respiratory rate of the central nervous system during development, shown in Fig. 5, appears to be in general agreement with the succinoxidase content.

The rate of enzyme accumulation of cholinesterase and succinoxidase in relation to total protein content in the central nervous system during the course of normal development is given in Fig. 6. The exponential function of the rate of increase in total protein content, measured as protein nitrogen, is characteristically a growth process generally described by the approximation $X = ae^{kt}$. The close parallelism of the synthesis of succinoxidase to the growth curve of the central nervous system

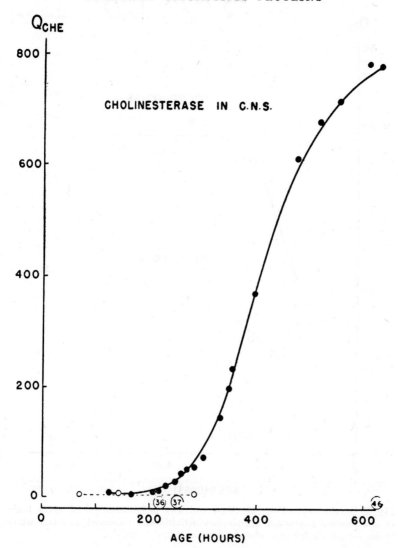

Fig. 4. Development of cholinesterase in homogenates of whole brain of *Amblystoma punctatum*. Developmental stages in encircled numbers. Open circles are enzyme determinations on samples without the addition of substrate acetylcholine. Q_{CHE} expressed as in Fig. 3.

is in sharp contrast to the excess and differential accumulation of cholinesterase during the same period of development. The relationship strongly favors the suggestion that, while the synthesis of respiratory enzymes in the central nervous system may be an index of general growth, the development of cholinesterase is significantly correlated with

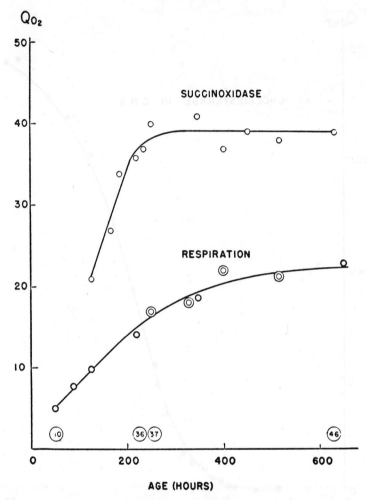

Fig. 5. Development of succinoxidase activities in homogenates of whole brain of *Amblystoma punctatum*. Respiratory activities are measured on intact whole brains (single circle) and subdivided brains (double circles).

the specific process of neural differentiation. Here is what seems to be a clear example of chemical heterauxesis of two enzyme systems of totally different developmental significance.

The intimate and specific association of the synthesis of cholinesterase but not of succinoxidase with the process of neural differentiation can be further demonstrated by experiments in which normal neural differentiation in the brain of an animal has been suppressed or modified. In amphibian brains an especially high concentration of cholinesterase is

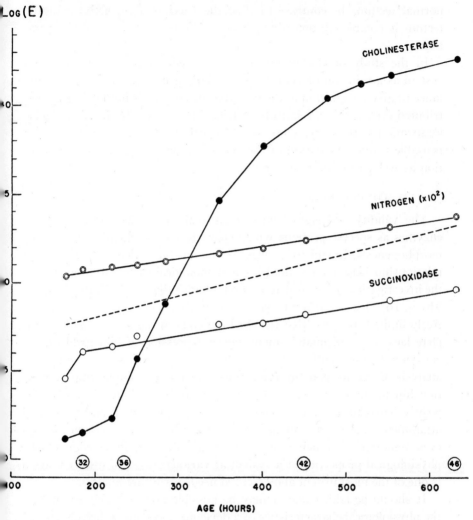

Fig. 6. Total enzyme content (E) of the central nervous system of *Amblystoma punctatum* at various stages of development. Broken line is the volume of the cns.

found in the fully differentiated optic tectum, among other centers (Shen, Greenfield, and Sippel, 1953). The development of the enzyme in this region is closely related to the functional differentiation of the optic system. When unilateral extirpation of the primordial eye vesicle is performed at an early stage, there is a marked suppression of the neural differentiation in the affected tectum (Larsell, 1929) accompanied by a significant deficiency in cholinesterase content as compared with the

[89]

normal tectum. In contrast to this, the succinoxidase activity in the tectum is completely unaffected by the defection in neural differentiation.

In the study of cholinesterase development in the central nervous system as a function of neural differentiation, it is of particular importance to give recognition to the fact that the enzyme is not uniformly distributed even within a relatively restricted area of the brain. The precise localization of the enzyme on a histological or cytological level is indispensable to the elucidation of the role of cholinesterase in neural function as well as in differentiation.

VII. CONCLUSIONS

The validity and significance of an analysis of the development of enzymes as specific proteins which characterize the chemical aspects of morphogenesis depend to a large extent on the evidence of a tangible relationship between the analysis and biological differentiation. From the investigations described above and other similar studies (Boell, 1948; Moog, 1952) it is apparent that results in this endeavor have been relatively limited. It seems that the chief difficulty at the present is the complete lack of an adequately quantitative expression of biological differentiation. Consequently, in spite of the relative precision in the chemical analysis of enzyme content, the results bear no precise quantitative relationship to functional or structural characteristics. Thus far, where a positive correlation is suggested between enzyme content and functional attainment in a given system, its nature has been necessarily crude and even tentative. A much more precise and detailed knowledge of the physiological properties of a system at various stages of differentiation is one of the most urgent and essential needs for future analysis.

It should be added that it may be possible and desirable to express the physiological characteristics of a developing system on broader terms than its specific function. It is well known that the total enzyme contents are generally in considerable excess to what may be considered as a physiological necessity under normal circumstances. In the nervous system, for example, it has been shown that a considerable portion of the enzyme cholinesterase may be inhibited by specific poisons without discernible interference with its normal function. This is particularly true in adult systems. It has often been suggested that the large reserve of enzymes in an adult system is probably of a homeostatic significance, as the adult organism is much more exposed to environmental exigencies than are embryos. If this view can be accepted, the "excess" accumulation of a specific enzyme at a given stage of development is not without its quantitative significance as a chemical index of functional differentiation and, perhaps more appropriately termed, capacity.

The present analysis of developmental enzymology, while conspicuously lacking in precision, may be regarded as the necessary beginning of a logical approach to chemical embryology. Being essentially exploratory in nature, these attempts are significant in revealing, among other things, the limitations of the present experimental technique and of interpretation. It is to be hoped that through such revelations the problem of chemical morphogenesis as a whole may perhaps be viewed in better perspective.

BIBLIOGRAPHY

Bailey, K. 1942. Myosin and adenosinetriphosphatase. *Biochem. J. 36,* 121-139.

Boell, E. J. 1948. Biochemical differentiation during amphibian development. *Ann. N.Y. Acad. Sci. 49,* 773-800.

Boell, E. J., and S. C. Shen. 1950. Development of cholinesterase in the central nervous system of Amblystoma punctatum. *J. Exp. Zool. 113,* 583-600.

Coghill, G. E. 1929. *Anatomy and the Problem of Behavior.* Cambridge University Press.

Detwiler, S. R. 1948. Quantitative studies on locomotor responses in Amblystoma larvae following surgical alterations in the nervous system. *Ann. N.Y. Acad. Sci. 49,* 834-855.

DeVillafranca, G. W. 1953. An investigation of the distribution and development of adenosinetriphosphatase activity in developing rat muscle. Doctor's thesis, Yale University.

Engelhardt, V. A., and M. N. Ljubimowa. 1939. Myosin and adenosinetriphosphatase. *Nature 144,* 668.

Herrmann, H., and J. S. Nicholas. 1948. Enzymatic liberation of inorganic phosphate from adenosinetriphosphate in developing rat muscle. *J. Exp. Zool. 107,* 177-182.

Herrmann, H., J. S. Nicholas, and M. E. Vosgian. 1949. Liberation of inorganic phosphate from adenosinetriphosphate by fractions derived from developing rat muscle. *Proc. Soc. Biol. Med. 72,* 454-457.

Ivanov, I. I., and B. S. Kasavina. 1948. Comparative biochemical study of the contractile proteins of the transversely striped muscle at various stages of phylo- and ontogenesis. *Doklady Akad. Nauk. U.S.S.R. 60,* 417-420.

Kielley, W. W., and O. Meyerhof. 1948. Studies on ATPase of muscle. II. A new Mg-activated adenosinetriphosphatase. *J. Biol. Chem. 176,* 591-601.

Larsell, O. 1929. The effect of experimental excision of one eye on the development of the optic lobe and opticus layer in larvae of the tree-frog (*Hyla regilla*). *J. Comp. Neur. 48*, 331-353.

Mihályi, E., and A. G. Szent-Györgyi. 1953. Trypsin digestion of muscle proteins. III. Adenosinetriphosphatase activity and actin-binding capacity of the digested myosin. *J. Biol. Chem. 201*, 211-219.

Moog, F. 1947. Adenylpyrophosphatase in brain, liver, heart and muscle of chick embryos and hatched chicks. *J. Exp. Zool. 105*, 209-220.

Moog, F. 1952. The differentiation of enzymes in relation to the functional activities of the developing embryo. *Ann. N.Y. Acad. Sci. 55*, 57-66.

Nachmansohn, D. 1939. Cholinestérase dans le système nerveux central. *Bull. Soc. Chim. Biol. 21*, 761-796.

Nachmansohn, D. 1950. *Nerve Impulse.* Josiah Macy, Jr., Foundation, New York.

Sacktor, B. 1953. Investigations on the mitochondria of the house fly, *Musca domestica L.* I. Adenosinetriphosphatases. *J. Gen. Physiol. 36*, 371-387.

Shen, S. C. 1949. Development of respiratory enzymes in rat muscle. *Anat. Rec. 105*, 489.

Shen, S. C., P. Greenfield, and T. Sippel. 1952. Application of histochemical technic for cholinesterase to paraffin sections. *Proc. Soc. Exp. Biol. Med. 81*, 452-455.

Szent-Györgyi, A. 1951. *Chemistry of Muscular Contraction.* Academic Press, New York.

V. CELLULAR DIFFERENTIATION AND THE DEVELOPMENT OF ENZYME PROTEINS IN PLANTS

BY R. BROWN[1] AND E. ROBINSON[2]

THIS CHAPTER is based on the results of a series of studies on the growth and consequent differentiation of the plant cell. The importance of studies on cell growth in general requires no emphasis. Ultimately the results of these are required in the interpretation of the differentiation and of the change in quantitative relations involved in the development of any multicellular organ. One aspect of the significance of observations with plant systems may, however, be emphasized. In many growing tissues cells divide sporadically throughout the population and at any one time the group contains cells in all stages of formation and development. It is only rarely that groups of cells can be isolated that are even approximately in the same stage of development. It is probably because of this that comparatively little information is available on the factors that control growth in the animal cell. The usual situation requires that observations on cell growth be restricted to individual cells and the exigencies of the techniques that can be applied necessarily limit the scope of the observations that can be made. Plants, however, present one of the rare situations in which groups of cells in different stages of development are spatially separated in the tissue. With this material the experimental situation is relatively simple and the variety of observations that can be made with it is comparatively large. It is therefore possible that data on cell growth of general biological significance can be more readily obtained with plant than with animal material.

The growth of a plant cell is a single, continuous, and consistent process. For analytical purposes, however, it is necessary to recognize in it three phases. It begins with the induction of growth, continues with the expansion of the system, and terminates with the cessation of growth. It is evident that the induction of a state from which growth can proceed is a phase which is independent of formation. The persistence of an active meristem indicates as much. The mere formation of a cell does not imply that growth will necessarily follow. When the state of potential growth is established a phase of expansion sets in which leads to

[1] Department of Agriculture, Oxford, England.
[2] Department of Botany, University of Leeds, England.

the state in which growth no longer occurs. This description, although no doubt trite, is nevertheless important since it defines the nature of the experimental approach appropriate to the situation. It is not enough to consider cell growth in the same terms as any other physiological process which has a ponderable product. The increment in dry weight that is involved in growth may be analyzed in the same general terms as, for instance, the dry weight change that occurs in a leaf as a result of photosynthesis. But it is important to notice that, whereas with the change in the leaf the system that promotes it remains constant, with the corresponding change in the growing tissue the system changes qualitatively during the period of observation. One of the essential characteristics of the cell that has started growing is that it is developing into a state in which further expansion does not occur. It seems to us probable that it is precisely these considerations that are implied by the plea of Erickson and Goddard (1952) for an emphasis on time in growth studies.

The hypothesis is proposed here that the growth of the cell is determined by a development which involves a succession of metabolic states. Before discussing the evidence on which this interpretation is based it may be appropriate to consider certain alternative hypotheses and the grounds on which they may be rejected.

The induction of a state of potential growth has not hitherto been recognized as one of the stages in the development of the cell, but at least two suggestions imply such a state. It has been suggested that cell growth is a process of inflation depending on the absorption of water. If true, this would involve the initial accumulation of a solute reserve. For the development of such a reserve there is no evidence and osmotic inflation is in any case improbable since it has been shown that the expansion of the cell is a true process of growth in the sense that at least in certain stages it probably involves increases in all components of the system. Another interpretation of the course of growth relies on the assumption of the utilization of a reserve of a substrate metabolite. This suggestion is discussed again below, but in this connection it may be noted that it also implies an initial phase in which the reserve is being supplied to the cell.

In the stage of active growth there is a considerable expansion and this undoubtedly depends on the whole metabolic activity of the cell. When aeration is restricted and in some cases when a nutrient supply is not available expansion is limited. The increase in volume is of course primarily due to the absorption of water, and the evidence indicates that when growth is proceeding the metabolic state is such that it promotes the processes on which the absorption of water depends. In time expansion ceases, and this might be due either to a change in the metabolic state or to one of the following:

(a) the attainment of the limit of elastic stretch in the wall of a cell which is being inflated by the metabolic absorption of water;

(b) the depletion of some substrate metabolite which is not synthesized by the cell but which is required and utilized in growth;

(c) the accumulation of an inhibitor during the course of growth;

(d) the establishment of some condition dependent on the geometry of the system.

It is probable that the absorption of water during growth is at least in part due to an active metabolic secretion of water into the vacuole (Thimann, 1951; Brown and Sutcliffe, 1950). Such being the case it is conceivable that the secretion might occur against an increasing back pressure of the wall and that after a certain critical stage has been reached this effectively prevents any further expansion. This is improbable, however, since the available evidence does not indicate that turgor pressure increases steadily as growth proceeds, and data on cellulose formation show that this continues after growth ceases (Burström, 1942; Brown and Sutcliffe, 1950).

If tissue that is potentially capable of growth is excised and supplied with water only for some hours before being transferred to a full nutrient medium, the final volume attained is less than it is when the tissue is transferred immediately to the full medium. This is taken to indicate that during the period on water some metabolite or metabolite complex which is required in growth has been consumed. It is argued that before growth begins a quantity of a hypothetical substance is accumulated and that when the original reserve has been depleted growth ceases. In this connection the effect of sugar on the growth of excised root tissue is of some significance. With increasing concentration of sugar not only the rate but also the time of growth is increased, and it is difficult to understand how sugar concentration could have this effect on time if the cessation of growth is being determined by the consumption of some other substance. Since the substance is not synthesized in growth and since sugar accelerates metabolic activity, two possibilities may be considered: rate of consumption of the hypothetical substance may depend on general metabolic activity, but in this case the time of growth should decrease with increasing concentration of sugar; the alternative possibility is that consumption of the substance is independent of metabolic activity, but here the time of growth should remain constant.

If cessation is due to the effect of an inhibitor and this inhibitor accumulates during the period of growth, then the rate of growth with excised tissues should be high initially and should decrease with time. Frequently this is the usual succession. In certain excised tissues, on the other hand, the rate initially tends to increase with time. This suggests that when a decreasing rate with time is observed this is not due to the operation of an inhibitor.

The geometry of the system may affect a variety of conditions. With cells from other organisms it has been suggested that as they expand the ratio of surface to volume decreases and that as a result the nutrient supply is depressed to the point at which it limits the reactions on which growth depends. Even if this interpretation is valid for the animal cell, it cannot be so for the plant cell. After a certain volume has been reached the cell grows as a cylinder and surface-volume relationships do not change substantially. Further it has been shown that during growth the thickness of the protoplast decreases and all parts of it therefore become more accessible to external nutrient supplies.

The changing geometrical pattern may of course also affect growth by influencing the internal distribution of cellular constituents. For instance when there is no increase in protein (as there is not in certain cases) the expansion of the cell necessarily involves a decrease in protein per unit surface. This decrease might continue to the point at which the distance between reactive centers becomes sufficiently great to limit critical metabolic processes. It is improbable, however, that this type of change is involved in the cessation of growth. If it were, then whatever the rate of expansion with different concentrations of a single nutrient, growth might be expected to continue until the limiting condition is established and in that event all the final volumes should be the same. In fact they are usually not the same. With different concentrations of sugar, for instance, the final volume tends to be smaller in correspondence with lower external concentrations.

All the above interpretations assume a possible stable metabolic state. Evidence from growth studies which has been reviewed elsewhere suggests that this is not the situation (Brown, Reith, and Robinson, 1952). Further evidence is presented here showing that there is a continued development in the metabolic pattern and that it is this that controls the growth process. The data given below represent the results of determination of two sets of characteristics which can be referred to the protein component of the cell. The one set of data shows changes in the total trichloracetic acid precipitable fraction, and the other, changes in certain enzyme activities which are taken as measures of qualitative changes in the protein complex and also of changes in metabolic activity. The activities that have been studied are those of an invertase, of an acid phosphatase, of a dipeptidase, of glycine oxidase, and of a proteinase system. It may be emphasized that the activities that have been examined have been chosen as indexes of general changes and not primarily because of any assumptions regarding their importance in the growth process. It is recognized that an exhaustive analysis of the developmental state of the protoplast can not depend on observations that may be referred to the protein component alone. Further it is clear that a comprehensive study of the protein complex itself requires an examination of the chemical and physical properties of different fractions. It is probable, however, that the data available, while fragmentary, are sufficiently extensive to justify conclusions as to the general nature of the system.

All the results discussed here have been obtained with the root, which in many respects is an admirable experimental object. The tip is occupied by a meristem from which cells are continuously being generated. Many of the cells formed become involved in growth, but as they do so they are replaced by others from the tip. Thus at increasing distances from the apex cells are in progressively advanced stages of development and the general technique used by us relies on this fact. Serial segments of tissue have been cut along the root from the apex basipetally throughout the region in which growth is occurring and the properties of these segments studied. The assumption is made that the cells in the segments are sufficiently uniform to justify assigning certain differences between the fragments to differences in the development of the cells of which they are composed. The justification for this assumption has been presented elsewhere (Brown, Reith, and Robinson, 1952).

The groups of segments are used in two different experimental situations. In the first the properties of the segments are analyzed immediately after separation from the parent root, in the second during culture after excision. In the first series of experiments after quantitative measurements are made with the tissue the results are related to the total number of cells it contains. This general technique yields a series of values which represent the changes that occur in the cell as it expands. The series of values obtained represent a temporal order. The time span involved in the change from one state to the next is not defined and rates for the changes observed can not therefore be calculated. Erickson and Goddard in a paper read to the Society for the Study of Development and Growth in 1951 described the results of some investigations also on cell growth in the root which involved a technique similar to that used by us. These workers, however, were able to elaborate the procedure to yield data that showed the rates at which the several changes they studied occurred.

In the second group of experiments the excised fragments are transferred to culture media and changes with respect to growth and other properties are studied during the period of culture and not immediately after excision. In this group of experiments the necessity to state results in terms of a unit cell does not arise since at least in those segments of the series in which growth occurs there is little change in cell number.

The two experimental series complement each other. Observations in the first show the changes that occur with growth in the intact root. Such observations characterize the nature of the process and define its full scope. Results obtained with excised fragments, on the other hand, provide the basis for analyzing the significance of the drifts found in intact tissue. The position in an intact organ is complicated by the fact that growth in these circumstances is influenced by a variety of unknown materials released from more and less mature tissues.

Most of the data described below have been obtained with the root of the broad bean and with successive 1.0 mm. segments taken over the first 10 to 15 mms. from the apex. Each experimental series therefore involves ten to fifteen groups of corresponding fragments. The techniques used in this series of studies have been described by Brown and Broadbent (1950) and by Robinson and Brown (1952). The numbers of cells in fragments have been determined by the maceration technique of Brown and Rickless (1949).

The enzyme activities have been determined with entire although dead tissue. Invertase activity has been measured with sucrose, phosphatase with glycerophosphate, dipeptidase with alanylglycine, glycine oxidase with glycine, and proteinase with hemoglobin. For activity determinations the tissue is killed at $-20°C.$, washed after thawing, and then transferred to the appropriate substrate. With this procedure it is possible that the rate of the reaction is limited by rates of diffusion to and from the reaction sites, but if any restraint is involved it affects all members of an experimental series and the relative values are therefore not likely to be modified. Further it has been shown (a) that with invertase and phosphatase activities these are greater after freezing than they are before it, and (b) that with phosphatase and proteinase the intact but frozen tissue gives results that are similar to those obtained with the most active extracts that have so far been prepared, and this is the case even when the substrate molecule is as large as a protein. The evidence suggests that the freezing and subsequent thawing tend to tear both the cytoplasm and the cell wall extensively enough to ensure relatively free access of substrate molecules to the reaction sites.

The enzyme data for the intact root are given below in terms of relative values per cell. These represent the reaction rates for the intact fragments divided by the number of cells and the smallest value of the series reduced to 1.0 with corresponding adjustments in all the others.

The main features of cell growth in the bean roots in this series of experiments are similar to those recorded for pea roots by Brown and Broadbent (1950) and for corn by Erickson and Goddard (1951). The water content data of Fig. 1 (which are a measure of volume) show that over about the first ten millimeters the volume of the cell increases about twentyfold. The older results show that this volume increase is accompanied by an increase in dry weight. These results also indicate that during growth protein content at first increases and then decreases until a constant steady level is established. The data of Brown and Broadbent suggested that the decrease in protein is coincident with the cessation of growth and those of Erickson and Goddard that it occurs before growth ceases. The data for the bean shown in Fig. 1 indicate that over about the first two millimeters it may decrease slightly, but that then it increases, and again decreases. The increase is about twofold, and the decrease occurs before overall growth ceases.

The changes in protein content are accompanied by changes in the activities of at least some enzymes in the cell. That such a correlation might be involved was originally suggested by certain observations on the respiration of developing cells. Brown and Broadbent showed that the increase in protein is accompanied by an increase in respiration per cell and similarly that the decrease involved a corresponding decrease in respiration. For various reasons it could be concluded that the changes in respiration were not due to variations in the concentration of substrate but that they were due to corresponding fluctuations in the levels of at least some respiratory enzymes. This interpretation is certainly consistent with the data of Figs. 2, 3, and 4, which show the levels of phosphatase, invertase, dipeptidase, and glycine oxidase activity per cell at increasing distances from the apex of the root of bean. With phosphatase (Fig. 2), after an initial slight decrease there is about a threefold in-

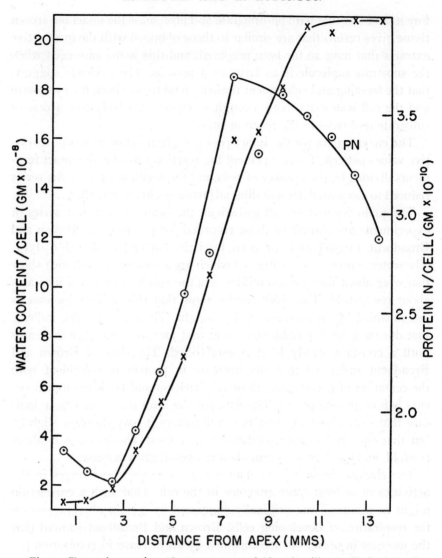

Fig. 1. Change in protein and water content (volume) with growth in cells of bean roots. Data represent values per cell at increasing distances from the apex of the root.

crease in activity which is then followed by a decrease. With the dipeptidase (Fig. 2) activity increases immediately from the tip, continues to increase until a peak value is reached, and then decreases. The changes with invertase (Fig. 3) and glycine oxidase (Fig. 4) are similar to those with phosphatase. It is significant that the changes in activity with all these enzymes tend to follow the changes in protein level. All with the

[100]

exception of the dipeptidase decrease as the protein decreases over about the first two millimeters, all increase when protein is increasing, all reach a peak value at approximately the same point as the protein, and all decrease with the protein. Clearly it is probable that the increment in protein represents at least partly an increase in enzyme protein, and similarly

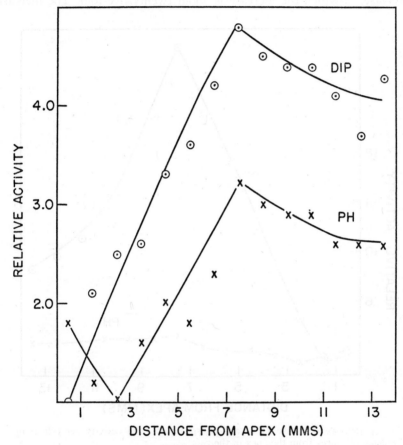

Fig. 2. Relative phosphatase (PH) and dipeptidase (DIP) activity per cell at increasing distances from the apex of the root.

that the decrease in the over-all fraction represents at least partly a dispersal of enzyme. It is also significant that, while there is close correspondence between enzyme activity and protein level, there is little apparent connection between protein and growth and between gross change in enzyme level and growth.

The activity data undoubtedly show that the over-all metabolic activity of the cell increases at least in the early stages of growth. However, they also probably indicate that the metabolic pattern changes. Fig. 2 shows that, whereas in the first eight millimeters the over-all increase for the dipeptidase is about fivefold, the corresponding increase for the phosphatase is only about double. In Fig. 3 the phosphatase values of Fig. 2 have been plotted on a different scale in order to emphasize the contrast with invertase, in which the increase is about twenty-five-fold. The increase

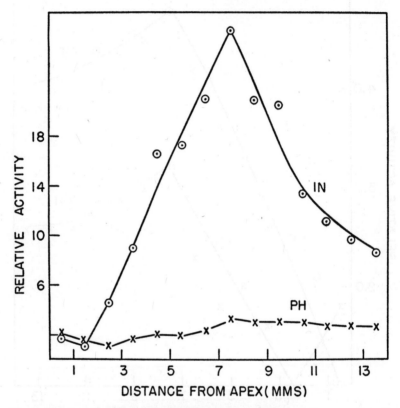

Fig. 3. Relative invertase (IN) and phosphatase (PH) activity per cell at increasing distances from the apex of the root.

in glycine oxidase shown in Fig. 4 is about fourfold and intermediate between the values for phosphatase and dipeptidase. These different relative changes suggest of course that the relative rates of different reactions change as the cell develops. They suggest that as the cell matures the metabolic pattern changes and that at different stages the metabolism of the cell may be dominated by different sets of reactions. Cell growth clearly involves a development in the sense that it is accompanied by a continuously changing metabolic state.

[102]

While the relative changes in activity promote differences in metabolic pattern, they are themselves probably the expression of variations in the composition of the protein complex. This aspect is emphasized by the curves of Fig. 5, in which the values of Figs. 2 and 3 are presented on a unit protein basis. These curves indicate that at least over the first six millimeters the proportion of invertase protein increases to a greater

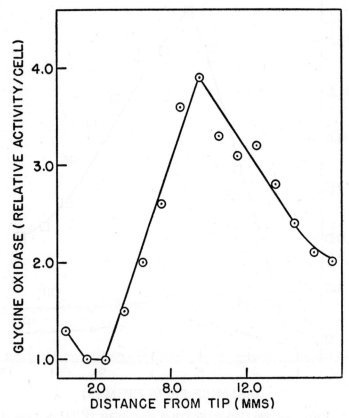

Fig. 4. Relative glycine oxidase activity at increasing distances from the apex of the root.

extent than does that of dipeptidase and to a greater extent again than that of phosphatase. While the proportion of invertase and dipeptidase is increasing, total protein is also rising and part of the relative increase in these enzymes may be due to a relatively greater synthesis of the proteins of which they are composed. At the same time part of it may also be due to a conversion of other proteins into these enzyme proteins during development.

[103]

The relative changes in enzyme activity might of course represent the effects of fluctuations in the levels of particular inhibitors and stimulators. This possibility is referred to again below, but in this context the results of certain preliminary observations on the amino acid composition of proteins from different parts of the root are relevant. Hydrolysates of

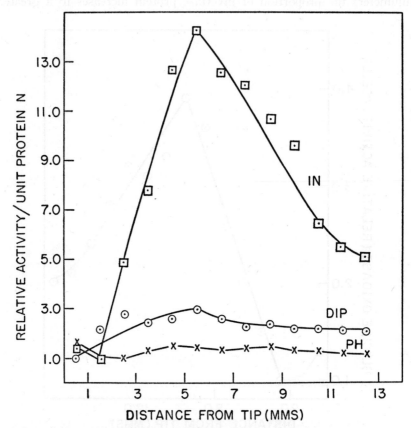

Fig. 5. Activity per unit protein nitrogen of phosphatase (PH), invertase (IN), and dipeptidase (DIP) at increasing distances from the apex of the root.

the proteins have been prepared and these have been examined chromatographically (Morgan and Reith, 1954). It has been found that at the apex of the root probably arginine and cystein or cystine are relatively abundant and that as growth proceeds these tend to decrease. Also several ninhydrin reacting substances may be present in hydrolysates from the apical region which are not as prominent in preparations from more mature regions. The evidence available, fragmentary as it is, is at least consistent with the position that the proteins do change and that the changes in the relative activities of the enzymes may be attributed to this.

The data of Fig. 5 are of further significance in another connection. Low activity per unit protein in the first two segments of the series is shown not only by the invertase, dipeptidase, and phosphatase data but also by the glycine oxidase values of Fig. 4 and by the proteinase data of Fig. 11. These results correspond to the earlier observation of Kopp (1948) that per unit protein respiration in the apex of the root is low. In the bean roots used in our experiments the first two millimeters were occupied mostly by strictly meristematic cells, which are completely un-differentiated and unequivocally embryonic. Clearly, in embryonic plant tissue much of the protein is relatively inert and metabolic activity in these cells is comparatively low. The intrinsic metabolic activity of the mature cell is evidently much higher than that of the meristematic cell. The contrary view has been widely expressed for a number of years, but this is based only on the fact that per unit volume of tissue respiration is several times greater with meristematic than it is with mature tissue. Meristems contain a large number of small nonvacuolated cells whereas mature fragments contain cells most of whose volume is due to the water in the central vacuole.

The analysis of the relation between growth on the one hand and protein content and metabolic states on the other has been developed in this series of studies through experiments with excised fragments in culture. This technique has been used for a number of years with a variety of tissues, having originally been developed by Bonner (1933) with coleoptile fragments. Brown and Sutcliffe (1950) examined the conditions in which it could be applied with root tissue. They reported that, when fragments are taken from the zone of 1.5 to 3.0 mms. from the tip of the root of corn and are transferred to 2 per cent sucrose containing some potassium chloride, they grow; within a period of about 36 hours they may extend to a length of about 6.0 mms. In the absence of sugar, growth is limited and there is no reaction to potassium. Also it was found that growth is very intimately dependent on the aeration that is provided in the culture. The results of these early experiments were of some interest since they showed that in root as in other plant tissues the expansion of the cell is a process that depends on metabolic states. This conclusion was reinforced by the observation that growth in these cir-cumstances also involves a considerable increase in dry matter which was attributed to a necessary synthesis of wall constituents. In its original form the technique depended on the assumption that the analysis of growth is accomplished most readily with tissues that expand rapidly

[105]

in culture. Accordingly only the one fragment taken from a restricted zone of the root was used. Recent experience, however, has suggested that the technique is capable of considerable elaboration. It is clear that comparative data obtained with tissues that do not grow or that only grow slowly as well as with others that grow rapidly provide the basis for more extensive interpretations than are possible with results based

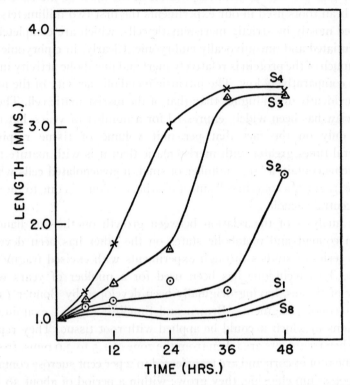

Fig. 6. Change in length with time in segments cultured on 2 per cent sucrose at 25°c. Segment S_1 occupies the zone in the parent root o to 1 mm. from the apex; S_2, 1 to 2 mms; S_3, 2 to 3 mms; S_4, 3 to 4 mms; S_8, 7 to 8 mms.

on tissues that at the time of excision are in the stage of development that will promote rapid growth. It is suggested that the general technique can be elaborated by the use of fragments that at the time of excision represent cells in various states of development.

The data of Fig. 6 show some of the results of a series of experiments in which the growth of successive segments on 2 per cent sucrose has been determined. The fragments are each 1.0 mm. in length and represent the same series as that on which the data for growth in the intact root were obtained. Thus at the beginning of the cultural period the successive fragments contain cells in progressively advanced stages of development. In the original experiments ten fragments were involved but in the interests of clarity in the diagram only the results with five are given. The results show that little or no growth is made by the first fragment during the cultural period of 48 hours. With each successive member of the series the increment is greater until with the fourth there is more than a threefold increase in length. With the fifth and subsequent members of the series the growth potential becomes progressively less until with the eighth, ninth, and tenth fragments growth in culture is negligible.

The course of growth with the earlier fragments is a feature of some importance. With the second segment growth is slow for the first 36 hours and then accelerates; with the third the lag period only lasts for 24 hours, and with the fourth rapid growth continues from the beginning of the experiment. The results suggest that in the intact root expansion is slow immediately after formation, that it accelerates with time, reaches a peak value, and then decreases. This inference has been confirmed by the observations of Erickson and Goddard (1952), who have shown in the root of corn that the rate of expansion increases with increasing distance from the apex and then decreases. Clearly, rapid growth is a consequence of a developmental process that has preceded it. Each fragment after excision traverses a development that is normally completed in the intact root. The second fragment at excision is composed of more developed cells than the first. Nevertheless some internal change has to occur in it before rapid growth can begin. The change occupies about 36 hours. Some of the development required in the second has already been completed in the third at the time of separation from the parent root and rapid growth can begin in it at 24 hours. With the fourth fragment the full development has already occurred at the time of excision and growth can begin immediately. In the first segment the developmental state is such that often no growth proceeds from it in culture.

The data of Fig. 6 sustain the view that the induction of the state in which growth can occur is a stage in the development of the cell. The induction of this state is clearly an autonomous process since it can be developed in culture when the nutritional conditions are certainly not as favorable as they are in the intact organ. The changes of course proceed more slowly in culture since the less elaborate nutrient environment necessarily imposes slower metabolic activity.

Whereas in the intact root, growth is accompanied by changes in the protein level, in the excised fragments whether growth occurs or not there is little change in protein content during the experimental period. The data for the third fragment are shown in Fig. 7. On 2 per cent sucrose protein content remains approximately constant. The data of Fig. 7 include a series obtained with tissue on water. When cultures are not supplied with sugar there is evidently a rapid decrease in the level of protein. It is significant, however, that even on water some growth occurs. Further it has been shown that increasing the protein level by culturing tissue on media containing a complex mixture of amino acids

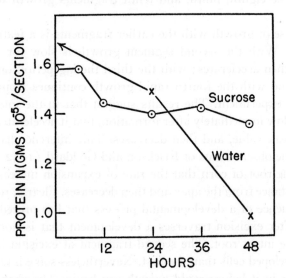

Fig. 7. Change in protein content in fragment S_3 of Fig. 6 on 2 per cent sucrose and water.

again has little influence on growth (Morgan, 1953). Indeed there is some evidence that in these circumstances it may be depressed. Clearly there is little or no connection between protein content and growth. This general conclusion is fully consistent with some of the features of the change in protein content in the intact root. In the first two millimeters while growth is proceeding protein per cell decreases and after the phase of increase protein starts decreasing before growth ceases.

The apparent independence of growth and protein content carries a number of implications. In particular it suggests that it may be a factor of some importance in determining the properties of the cell when growth has ceased, that it may in other words be one of the conditions through which the physiological differentiation of the cell is determined.

The extent to which protein is dispersed in the cell might not be important if the quantity in it is large. It might be important however if the quantity is so small that slight variations in its distribution affect the physical properties of the protoplast. There is some evidence that the protein content is in fact of this order. An estimate of the relative magnitude of the protein content may be made by calculating the amount required to form a single monolayer within the surface of an average vacuolated cell and then determining the number of such layers that can be provided from the protein available. The figure obtained constitutes a rough estimate of the number of protein molecules that are dispersed in a line across the protoplast. The position may be considered in terms of a cell 150μ long by 50μ broad and carrying 2.0×10^{-10} gms. protein N per cell. If it is assumed that the average molecular weight of the amino acids is 150 and the average area occupied by each amino acid is 20 sq. Å., then the number of protein layers across the protoplast is about 35. The figure for the amount of protein N per cell is a compromise between the values of Robinson and Brown and the slightly lower ones of Erickson and Goddard. The estimated value of 35 takes no account of protein in nuclei, mitochondria, and other cellular inclusions.

The calculated figure of course only indicates that the probable value is something between 10 and 100. It is an astonishingly small figure and quite certainly indicates that the position can not be considered in terms of thousands or even of hundreds of layers. The implications of this conclusion are considerable for the interpretation of the structure of the protoplast. Of immediate relevance is the possibility that whereas the reduction in the number of layers from $5,000$ to $2,500$ may not have very much effect on, for instance, the physical coherence of a hypothetical system, a corresponding reduction from 50 to 25 is likely to influence the position very profoundly and slight variations in the growth of the cell are therefore likely to modify the final properties of the cell considerably. Direct experimental evidence that this is the case is not available, but the results of certain observations on the absorption of potassium by expanding cells are at least compatible with this interpretation. Serial fragments each 1.5 mms. in length were taken from the tip along the root of corn and the absorption of potassium by each fragment examined. The results for the second and fourth segments are shown in Fig. 8, in which the calculated concentration values are also presented. The rate of absorption is greater with the second, but concentrations in the second are considerably lower than they are in the fourth. The second fragment contains more cells and the higher rate of absorption must be attributed to this condition. The second, however, grows in culture and the fourth does not, and as a result the cells of the second reach ultimately approximately the same size as those of the fourth but have a considerably lower protein content. The result of this is the development of a state in

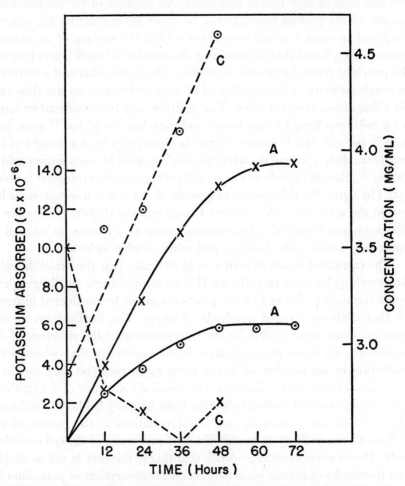

Fig. 8. Absorption (A) and change in internal concentration of potassium with time (C) in fragments from corn roots taken from zones 1.5 to 3.0 mms. (crosses) and 4.5 to 6.0 mms. from apex (circles).

which the permeability to potassium is greater, the rate of outward diffusion higher, and the final internal equilibrium concentration of potassium consequently lower (Brown and Cartwright, 1953).

While the growth of the cell is probably not controlled uniquely by the content of protein, it is probably profoundly influenced by qualitative

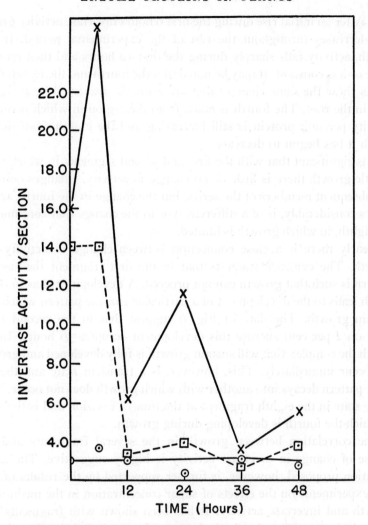

Fig. 9. Change in invertase activity with time on 2 per cent sucrose with segments S_1 (indicated by circles); S_4 (indicated by crosses); and S_8 (indicated by squares) of Fig. 6.

changes in this complex. Determinations have been made of invertase and phosphatase activity during culture with each segment of the bean series. Changes have been recorded in some of these in circumstances in which, be it emphasized, there is no change in protein. In the third and subsequent fragments of the group phosphatase tends to increase slightly with time. The results obtained with three of the fragments with invertase are shown in Fig. 9. It is evident that again the first fragment shows little or no change. In the fourth the change is consider-

[111]

able. After an initial rise during the first 6 hours invertase activity gradually decreases throughout the rest of the experimental period. In the eighth, activity falls sharply during the first 12 hours and then remains more or less constant. It may be noted that the fourth and the eighth fragments show the same changes that are normally traversed by these tissues in the root. The fourth is taken from the region in which invertase activity per unit protein is still increasing and the eighth from that in which it has begun to decrease.

It is significant that with the first and second segments in which there is little growth there is little or no change in activity. Changes occur in all subsequent members of the series, but the change in the fourth, which grows considerably, is of a different type to the change that continues in the eighth, in which growth is limited.

Clearly there is a close connection between change in activity and growth. The evidence suggests that in the first fragment the enzyme pattern is such that growth can not proceed. A development has to occur which leads to the development of a particular enzyme pattern which will sustain growth. The data of Fig. 6 suggest that in the second fragment on 2 per cent sucrose this development occupies 36 hours. In the fourth the complex that will sustain growth is fully developed and growth can occur immediately. This, however, is a transient state and the enzyme pattern decays into another with which growth does not occur. This is the state in the eighth fragment at the time of excision and is the state to which the fourth is developing during growth.

The correlation between growth in the several fragments and the course of change in invertase activity is highly suggestive. The interpretation proposed, however, is further supported by the results of certain experiments on the effects of sugar concentration in the medium on growth and invertase activity. It has been shown with fragments that expand after excision that both the rate of growth and the time during which growth continues increase with increasing concentration of sugar in the culture medium. In terms of the interpretation proposed above, in a fragment which grows rapidly as soon as it is excised the cells carry a full complement of growth enzymes, and as growth proceeds these decay and are converted into some other form of protein. With increasing substrate sugar the activity of these enzymes increases and growth is therefore accelerated. At the same time the fact that sugar increases the time of growth would suggest that sugar depresses the rate of enzyme decay. In this connection the effect of sugar on invertase decay is suggestive. Fragments which grow were cultured on different concentrations of sucrose and the invertase activity determined after 48 hours with each treatment. The results are shown in Fig. 10. With increasing concentration of sucrose the final invertase activity tends to increase. Evidently sugar tends to depress the rate of invertase decay. The implication of this is not that invertase is particularly important in growth (although it may be); the result does, however, suggest the possibility (1) that some other enzyme complex which catalyzes reactions important in growth, and in which sugar or products of sugar are involved, also decays during growth, and (2) that the substrate tends to protect the enzymes against breakdown and hence the effect of sugar in prolonging the time of growth.

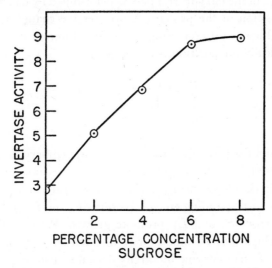

Fig. 10. Effect of sucrose concentration in the medium on final invertase activity in segments cultured at 25°C.

The change from one enzyme state to the next may again be due to the operation of a complex set of inhibitors and stimulators. While with excised fragments there is little or no direct evidence available for a change in the composition of the protein during growth, the indirect evidence is suggestive. Hydrolysates of the protein of fragments which grow vigorously have been analyzed chromatographically and the results suggest that slight changes may occur. The evidence from this series of observations, however, is inconclusive. At the same time as shown above some of the enzyme changes in excised fragments are of the same character as those that proceed in the intact root, and it is clearly improbable that the same inhibitors and stimulators would be produced from the complex nutrient supply available to the cell in the root and from the simple nutrients provided in culture. Thus the position may be accepted at least tentatively that in the fragments as in the intact root the change in enzyme activity is due to change in the state of the protein.

The interpretation of cell growth in the intact organ may now be summarized. It is determined by a development in the metabolic state. Immediately after the cell is formed the metabolic state is such that growth can not proceed from it. The cell, however, develops into another state which does support growth, and finally into one in which growth ceases. The several states are characterized each by a particular enzyme complex which is different from that of the succeeding and preceding states. Finally change in the enzyme complex is determined by a primary change in the composition of the protein. A progressive change is involved in the protein which at different stages sustains different enzyme patterns. Diagrammatically the position is as follows:

$$P_1 \rightarrow P_2 \rightarrow P_3 \rightarrow P_4$$
$$E_1 \quad E_2 \quad E_3 \quad E_4$$

P_1 represents the protein of the meristematic embryonic state and E_1 the enzyme complex corresponding to it. E_1 does not sustain growth but the protein develops into P_2 and in time into the state of P_3. P_2 and P_3 involve the corresponding enzyme patterns E_2 and E_3 and these, which are different from E_1 and E_4, promote growth. With further development P_3 develops into P_4 and as a result another enzyme complex E_4 emerges. This fourth enzyme hierarchy does not sustain growth and the development from P_3 and P_4 therefore involves the cessation of growth.

Once the change from P_1 to P_2 has occurred growth will be affected by two conditions: by the rates of the reactions catalyzed by E_2 and E_3, and by the rates of the changes from P_2 to P_3 and from P_3 to P_4. The rates of the reactions involved with E_2 and E_3 will vary with temperature, substrate concentration, and so on, and the higher these rates the higher the rate of growth. The rate of the change from P_2 to P_4 will of course control the length of time during which growth proceeds. In this phase the slower the reactions involved the longer is growth likely to continue and consequently the larger the final cell volume is likely to be.

Clearly the stimulation of growth involves as much the inhibition or retardation of certain reactions as it does the acceleration of others. It has frequently been observed that metabolic inhibitors at low concentrations tend to stimulate growth. It is possible that the stimulating effect of substances such as iodoacetate and protoanemonin is due to the retardation of the developmental changes in the protein.

The discussion of change in enzyme level has been framed in terms which suggest that it is change in the absolute level that controls growth. This is undoubtedly the position in excised fragments. On the other hand if it is simply a matter of absolute level, then change in protein content might be expected to influence growth. But change in the absolute level of one enzyme, even when protein is constant, also represents a change relative to other enzymes, and it is this which is evidently the dominating feature. In the intact root change in protein as such will affect the position only insofar as it modifies the relation between one enzyme level and another.

In the isolated fragment protein does not increase since a supply of amino acids is not available to it. In the intact root this supply is continuous and the content can increase. At the same time this is not the only factor involved. It is possible that the changes in total protein are themselves in part a consequence of the change in the composition of the protein.

It is frequently assumed that change in protein content is a simple expression of synthesis. Such it almost certainly is not. Increase in protein is probably a consequence of a relatively higher rate of synthesis than of degradation and decrease the converse state of a relatively higher rate of degradation. In the same sense a constant protein content probably represents a steady state between synthesis and degradation. Unfortunately neither phase can be attributed to particular catalytic systems. On the other hand it is possible that proteolytic enzymes are involved in the degradation process, and they may be the agents through which the turnover in the protein complex is maintained. The data of Fig. 11 (Robinson, 1953) may be considered with this possibility in view. These data show the change in proteolytic activity per cell from the meristematic to the fully developed state of the fifteenth segment. The position with the proteinases is evidently similar to that recorded with other enzymes—although there are important differences. As with other enzymes proteinase activity increases at first and then decreases. Moreover in the meristematic region again activity per unit protein is low and increases with increasing distance from the apex. It is a possibility that in meristematic and associated tissues protein increases as a result of low proteolytic activity. The results obtained with the first two fragments in culture are consistent with this interpretation. Enzyme activity remains relatively stable, suggesting that degradation is slow. As

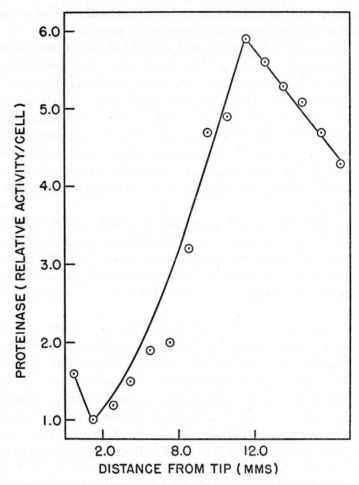

Fig. 11. Relative proteinase activity at increasing distances from the apex of the root.

development proceeds proteolytic activity becomes relatively greater and the difference between synthesis and degradation decreases. At a certain point the rates of the two phases become equal. This probably occurs at about eight millimeters, and if there were no further change after this point the protein content might be expected to remain constant. The data suggest the possibility, however, that proteinase activity continues to increase, hence that after this point degradation exceeds synthesis and that it is as a result of this state that protein starts decreasing. The decrease in protein might be expected ultimately to affect the proteolytic enzymes themselves, and the decrease in proteinase activity which occurs at about ten millimeters might be interpreted as an effect of the proteinase on the protein of which it is composed. The decrease in proteinase activity thus occasioned might be expected to lead to a steady state between synthesis and degradation with the consequent establishment of a constant protein content.

The absolute and relative increase in proteinase must of course be attributed to the same condition as that which promotes the corresponding changes with other enzymes. No suggestion can be offered as to what the mechanism is which determines the change in the composition of the protein. One final point, however, may be noted. The fact that it occurs is a matter of some significance in relation to differentiation in general. It has been shown that the development in the protein pattern controls growth and growth is indeed one form of differentiation. It has been shown how the progress of growth with different protein contents may affect the physiological properties of the cell, and this again is a form of differentiation. But it is clear that other implications again are involved in the situation. Each of the protein states characterized above as P_1, P_2, P_3, and P_4 carry a particular enzyme complex. It is probable, however, that the complex characteristic of each state may be subject to some variation. If this is the case, then through the same fundamental pattern of changes cells of different physiological and therefore of morphological type may be developed.

BIBLIOGRAPHY

Bonner, J. 1933. The action of the plant growth hormone. *J. Gen. Physiol.* *17*, 63.

Brown, R., and P. Rickless. 1949. A new method for the study of cell division and cell extension with preliminary observations on the effect of temperature and nutrients. *Proc. Roy. Soc. (B)*, *136*, 110.

Brown, R., and J. F. Sutcliffe. 1950. The effects of sugar and potassium on extension growth in the root. *J. Exp. Bot. 1*, 88.

Brown, R., and D. Broadbent. 1951. The development of cells in the growing zones of the root. *J. Exp. Bot. 1*, 247.

Brown, R., W. S. Reith, and E. Robinson. 1952. The mechanism of plant cell growth. *Symp. Soc. Exp. Biol. 6*, 329.

Brown, R., and P. M. Cartwright. 1953. The absorption of potassium by cells in the apex of the root. *J. Exp. Bot. 4*, 197.

Burström, H. 1942. Die osmotischen Verhältnisse während des Streckungswachstums der Wurzel. *Ann. Agric. Coll. Sweden 10*, 1.

Erickson, O. R., and D. R. Goddard. 1951. An analysis of root growth in cellular and biochemical terms. *Growth*, Suppl. Vol. XV, 89.

Kopp, H. 1948. Über das Sauerstoff bedurfnis wachsender Pflanzenzellen. *Ber. Schweiz. bot. Ges. 58, 283.*

Morgan, C. 1953. Nitrogenous constituents of growing cells of the root. Doctor's thesis, University of Leeds.

Morgan, C., and W. S. Reith. 1954. The composition and quantitative relations of protein and related fractions in developing root cells. *J. Exp. Bot. 5,* 119.

Robinson, E., and R. Brown. 1952. The development of the enzyme complement in growing root cells. *J. Exp. Bot. 3,* 356.

Robinson, E., and R. Brown. 1954. Enzyme changes in excised root fragments. *J. Exp. Bot. 5,* 71.

Robinson, E. 1953. Unpublished data.

Thimann, K. V. 1951. Studies on the physiology of cell enlargement. *Growth,* Suppl. Vol. XV, 5.

VI. SOME PROBLEMS OF SPECIFICITY IN THE SEXUALITY OF PLANTS

BY JOHN R. RAPER[1]

I. INTRODUCTION

SPECIFICITY in reproductive processes of plants is achieved through the operation of many different basic mechanisms, each of which may be variously modified in its expression. Of necessity specificity is imposed at many different levels of the reproductive progression and has been shown in a few cases to constitute a major factor in the determination of the sequence of events in the reproductive process (Moewus, 1951; Raper, 1951). Primary emphasis in the present discussion, however, will be placed on the role of specificity in limiting sexual interaction among individuals of the same species or of closely related species and on the mechanistic aspects of such limitation.

Obligatory cross-mating, essential for the attainment of maximal benefit from genetic recombination, occurs in all major groups of plants and is imposed by numerous specific genetic devices. Certain of these devices are widespread throughout the plant kingdom while others are restricted to relatively small subgroups of plants; similarly, at the level of subdivisions different groups vary appreciably in their versatility as reflected in the range of restricting patterns which they exhibit: obligatory cross-mating is not known among the extant terrestrial ferns while no less than seven distinct patterns imposing cross-mating are known among the fungi (Raper, 1954). With only a few exceptions all of the restrictive devices can be divided into two major categories: those in which sexual factors are determining, and those in which nonsexual or extra-sexual factors, commonly termed incompatibility factors, are determining. The few exceptional cases in which determination is shared by sexual *and* incompatibility factors happily need not concern us here.

The genetic characteristics of the major cross-mating patterns are listed in Table I, as adapted from Whitehouse (1950); for each pattern is also given the group or groups of plants in which it occurs and its relative effectiveness in restricting inbreeding and favoring outbreeding. It is of particular interest that mating control by incompatibility factors

[1] Department of Botany, University of Chicago. The preparation of this chapter was materially aided by a grant from the Dr. Wallace C. and Clara A. Abbott Fund of the University of Chicago. Present address of the author is: Biological Laboratories, Harvard University.

[119]

occurs only in two groups, the fungi and the flowering plants, and it is felt by certain authors that the role of these factors in the evolutionary histories of these groups has been of extreme importance (Mather, 1944; Whitehouse, 1950).

TABLE I. The major genetic devices which impose mating specificity in plants.

	Determining factors	Effectiveness		Occurrence
		Inbreed $-\%$	Outbreed $-\%$	
A. SEXUAL FACTORS				
Haploid	X vs Y	50	50	Algae, Phycomycetes, Bryophytes
Diploid	XX vs XY	0	50	Gymnosperms, Angiosperms
B. INCOMPATIBILITY FACTORS				
I. 2-allele, I-locus				
Haploid	A vs a	50	50	Ascomycetes, Rusts, Smuts
Diploid	Ss vs ss	0	50	Angiosperms
2. 2-allele, 2-locus				
Diploid	$SsKk$ vs $sskk$	0	66 *	Angiosperms
3. Multiple-allele, I-locus				
Haploid	A^x vs A^y	50	ca. 100	Hymenomycetes
Diploid	S^xS^y vs S^mS^n	0	ca. 100	Angiosperms
4. Multiple-allele, 2-locus				
Haploid	A^xB^x vs A^yB^y	25	ca. 100	Gasteromycetes, Hymenomycetes

* One dominant allele is epistatic to the other, resulting in three self-sterile and cross-fertile classes.

The variety of basic mechanisms which ultimately determine self-sterility and cross-fertility would suggest an equal variety of specific agents operative in achieving the terminal effect. The fundamental dissimilarities evident in the few detailed and integrated mechanisms which have to date been demonstrated support this view, but too few restrictive mechanisms have been adequately elucidated to provide sufficient information for any inclusive and meaningful comparative synthesis. Exploration of a few selected restrictive mechanisms, however, will serve both to illustrate certain basic principles whereby specificity in reproduction is achieved, and to emphasize the limitations of current detailed information in regard to this important function.

The examples chosen for this discussion comprise an aquatic fungus and a flowering plant in which cross-mating specificity is achieved by sexual hormones and specific inhibitors, respectively, and a woodrotting mushroom in which an intriguing complex of distinct mating specificities, the responsible terminal agents for which are unknown, can be correlated directly with the determining genetic device.

II. MATING SPECIFICITIES AND SEXUAL HORMONES

Among the aquatic Phycomycetes, a group which is commonly considered to be primitive, a majority of the species are homothallic, i.e. hermaphroditic and self-fertile; however, there are many heterothallic species, in each of which are two or more self-sterile and cross-fertile strains. Self-sterility and cross-fertility in these forms result from the sexual differentiation of haploid individuals by the meiotic segregation of sexual factors. The details of the sexual progression have been determined in a number of heterothallic and homothallic species of *Achlya,* a genus of "water-molds," and certain of the interactions demonstrated in these forms have particular pertinence in the present discussion.

The entire sexual process, with the exception of the physical transfer of ♂ nuclei in fertilization, has been shown to be coordinated by a series of specific sexual secretions (see papers by Raper from 1939 to 1950). The sexual process is composed of a number of distinct stages which occur alternately in the ♂ and in the ♀ ; each stage is characterized by one or more morphological developments and is directly dependent both for initiation and for quantitative regulation upon the secretion(s) of the last preceeding stage. That the correlative agents operative here are indeed specific secretions has been amply demonstrated by their activity through permeable membranes, in cell-free filtrates, etc. The hormonal mechanism as it operates under normal circumstances, i.e. between ♂ and ♀ individuals of the same species in physical contact, is schematically diagrammed in Fig. 1.

Four distinct hormones, two from the vegetative ♀ mycelium and two from the vegetative ♂ and collectively termed the A-complex, initiate the entire sexual progression by stimulating the ♂ plant to produce antheridial hyphae, the primordia of ♂ sexual organs, which, originating as tiny lateral bumps upon the vegetative hyphae, rapidly elongate into branched sinuous filaments. The relative number of antheridial hyphae and the extent of their growth are determined by the quantities of the hormones of the A-complex which are present in the surrounding medium (Raper, 1942; 1950).

[121]

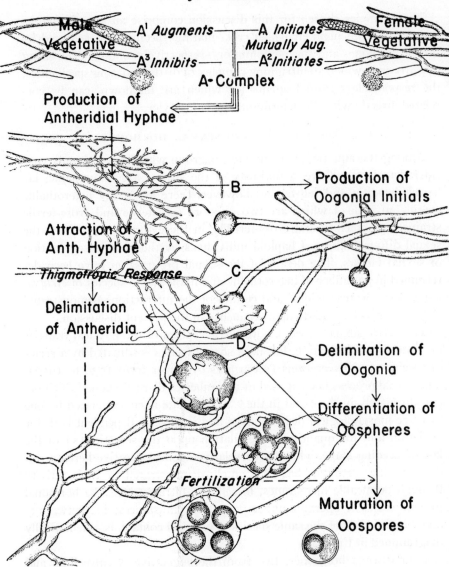

Fig. 1. A semidiagrammatic representation of the sexual progression in heterothallic species of *Achlya* relating the sequence of morphological developments to the origins and specific activities of the several sexual hormones.

Once the antheridial hyphae become established they secrete into the medium a diffusable substance, hormone B, which stimulates the ♀ plant to produce oogonial initials, the primordia of ♀ sexual organs. These primordia, originating as large lateral bumps upon the vegetative hyphae, elongate briefly prior to the production of large densely filled spheres at their tips.

[122]

The oogonial initials, during a period of a few hours to about a day following their maturation, secrete another agent, hormone C, which has two distinct effects upon the antheridial hyphae of the ♂ : (a) it effects the oriented growth of the antheridial hyphae toward the source of the secretion, the oogonial initial, and (b) it causes the antheridial hypha, after it has reached the oogonial initial and has become appressed to the wall of the ♀ structure, to delimit at its tip a unicellular multinucleate antheridium, the ♂ gametangium. That hormone C actually comprises two or more specific substances is suggested by the two distinct responses which it induces.

The final stage known to be hormonally controlled is initiated and regulated by an agent, hormone D, secreted by the fully differentiated antheridium. Here, as in the case of hormone C above, two distinct effects obtain: (a) the delimitation by septation of the terminal spherical oogonium, the ♀ gametangium, and (b) the reorganization of the dense peripheral layer of oogonial protoplasm to produce a small number (4 to 15 in these species) of spherical uninucleate oospheres, the ♀ gametes. Again a hormonal complex is suggested by the duality of physiological activity.

Fertilization is accomplished by the growth of tiny tubes from the antheridium into the lumen of the oogonium and the discharge of a single ♂ nucleus in each ♀ gamete. The activity of chemical agents in the regulation of these final events of the sexual process is suspected, but no experimental means have yet been devised to test this possibility.

In crosses between fully compatible plants the sexual progression can be experimentally interrupted at practically all stages which are hormonally initiated. One example, observed in membrane matings of *A. bisexualis* ♀ × ♂, will serve to illustrate a general type of disruption which can be specifically applied at various points in the correlative mechanism. When fully differentiated ♂ sexual organs, antheridia, are physically separated by means of a permeable membrane from ripened oogonial initials, the subsequent behavior of the latter depends upon their size, their distance from the membrane barrier, and upon the concentration of ♂ organs appressed upon the membrane. Within a critical distance, probably reflecting the limit of an effective concentration of diffused hormone C, small oogonial initials undergo the normal developmental processes of delimitation and differentiation of oospheres; large oogonial initials lying near the membrane and all oogonial initials lying beyond a critical distance, however, either enlarge and eventually degenerate, produce successive crops of new initials by proliferation until exhausted, or proliferate into normal vegetative hyphae. At this and at other stages the normal progression thus depends upon specific responses of specifically differentiated organs to specific chemical agents.

Specificity of hormonally regulated processes in these plants is even more strikingly shown by the reactions between individual vegetative plants in interspecific and even intergeneric matings. In numerous homothallic species of *Achlya* and the closely related genus *Thraustotheca* there has been demonstrated a hormonal coordinating mechanism similar to but somewhat simpler than that of the heterothallic species (Raper, 1950). Numerous interspecific matings between heterothallic species and between heterothallic and homothallic species have revealed a wide range of hormone-response specificities and the interactions in such contrasts run the gamut from mutual and total indifference to complete compatibility, i.e. sexual consumation. The most highly advanced sexual stage attained in each of a number of interspecific contrasts in respect to the over-all hormonal mechanism is shown in Fig. 2. The varying degrees of interaction displayed here can most feasibly be interpreted as reflecting —except in the occasional cases of complete compatibility—the failure of the various systems to provide, at the critical instances, initiating hormones of the specifically required chemical configurations.

Thus in the heterothallic members of this group, hormonal specificities operate to impose a pattern of self-sterility and cross-fertility. The restrictions imposed, however, are not species-specific. Interspecific reactions accordingly occur in many cases, but these in turn may be blocked at intermediate stages by hormonal specificities.

The fuller elucidation of the hormonal mechanism and the sexual reactions which it regulates has been seriously hampered by (a) the lack of exact chemical information about the hormones and the difficulties attendant to its achievement and (b) the inability to investigate the genetic basis of the mechanism because of the absolute reluctance to date of the zygotes of heterothallic species to germinate in the laboratory.

III. POLLEN DEVELOPMENT AND SPECIFIC INHIBITORS, PHYSIOLOGICAL DIFFERENTIATION, ETC.

A majority of the flowering plants, the Angiosperms, are hermaphroditic, with the ♂ and ♀ sexual organs, stamens and pistils respectively, commonly occurring in the same "perfect" flowers. Obligatory cross-breeding is imposed in an appreciable fraction of such hermaphroditic species by incompatibility mechanisms which, in certain combinations, prevent the germination of the pollen grain upon the stigma or the growth of the pollen tube through the style (see Lewis, 1949, for review).

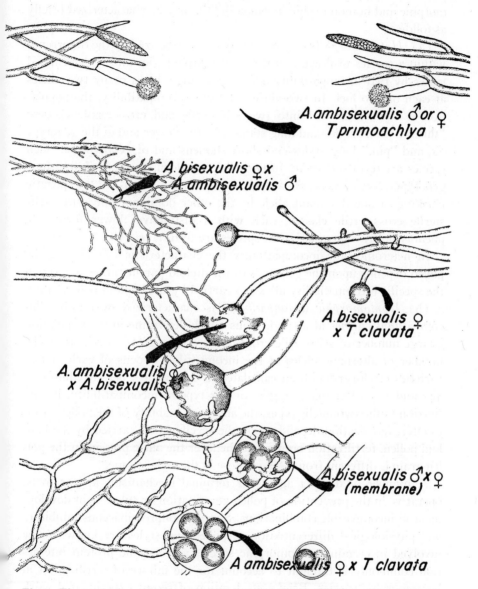

Fig. 2. The most advanced stage of sexual interaction attained in each of a number of interspecific and intergeneric matings between heterothallic and homothallic individuals. The hormonal mechanisms of the two mates are of necessity completely interactive, in each case, at all stages prior to the point of disruption due to hormonal specificity.

The basic types of incompatibility control in the Angiosperms, heteromorphic and homomorphic incompatibility, may be characterized briefly as follows:

(a) *Heteromorphic incompatibility* is characterized by the occurrence within a species of flowers of different morphologies and is determined either by two incompatibility alleles at a single locus or by two alleles at each of two loci. In two-allele one-locus incompatibility, the population is equally divided into two self-sterile and cross-fertile classes: "thrum," short styles and long stamens in the flower and of the genotype *Ss*, and "pin," long styles and short stamens and of genotype *ss*. Such species are termed *distylic*. In two-allele two-locus incompatibility, four genotypes, *SsKk, Sskk, ssKk,* and *sskk*, occur but because of the epistatic masking of one dominant allele by the other dominant only three self-sterile cross-fertile classes, each with a distinctive flower form, are phenotypically expressed. Such species are termed *tristylic*.

In heteromorphic incompatibility the characteristic behavior of the pollen depends upon the genotype of the diploid parent rather than upon the specific incompatibility allele(s) carried by the haploid pollen grain.

(b) *Homomorphic incompatibility* is characterized by multiple incompatibility alleles at a single locus which determine in the population a large number of self-sterile cross-fertile classes, S^1S^2, S^3S^4, etc. (The number of alternate alleles at the incompatibility locus of each of two varieties of clover has been calculated as 212 with 5 per cent limits of 442 and 115 (Bateman, 1947). In this type of incompatibility, in distinction to the two-allelic types, the ability or inability of the pollen grain to effect fertilization depends upon the specific allele carried by the haploid pollen, fertilization occurring whenever the allele present in the pollen grain is different from both of those in the stylar tissue.

A number of distinct agents or terminal mechanisms appear to be operative in the prevention of pollen germination or subsequent development in incompatible combinations. Specific nonproteinaceous inhibitors and physiological differentiation of pollen and styles are known to be involved in certain heteromorphic species and specific protein interaction commonly is thought to account for the failure of fertilization in homomorphic species. Brief consideration of recent experimental work in a few well-known species will serve both to illustrate the basic phenomena of specificity control in the flowering plants and to emphasize the mechanical variations by which specificity is imposed in different cases.

The well-known ornamental shrub *Forsythia intermedia* is a typical distylic species in which pollen germination and fertilization follow upon the transfer of pollen from long stamens to long styles, i.e. *pin* (style) × *thrum* (pollen), and pollen from short stamens to short styles, i.e. *thrum* (style) × *pin* (pollen). The biochemical details of the mechanism which provides this cross-breeding specificity have recently been revealed in investigations by the German biologist Moewus (1950).

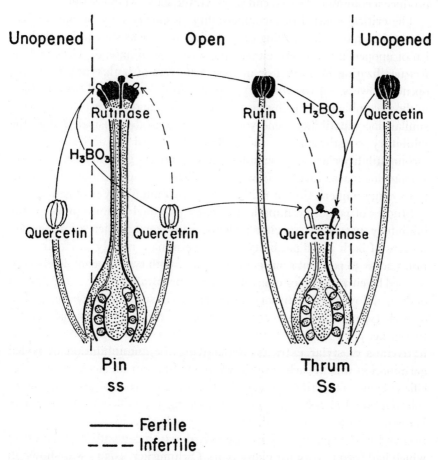

Fig. 3. Diagram of the mechanism of control of pollen-style incompatibility in *Forsythia intermedia* [summarizing data of Moewus (11)].

Pollen from each of the two types of flowers (Fig. 3) was shown to contain a specific inhibitor of its own germination, in each case a glycoside of the flavonol *quercetin:* pollen from long stamens (*Ss, thrum*) contains *rutin,* the rutinose (glucose + rhamnose) glycoside of quercetin, while the pollen from short stamens (*ss, pin*) contains the corresponding rhamnose glycoside *quercetrin*. It was further demonstrated that the styles of the two types contain specific enzymes which preferentially hydrolyze the two flavonol-glycoside inhibitors: short styles (*Ss*) contain a quercetrin hydrolyzing enzyme and long styles (*ss*) contain a rutin hydrolyzing enzyme. The activities of the two enzymes are absolutely specific in respect to the hydrolysis of the two glycosides but both produce a common inactive end product, the flavonol quercetin.

The critical control of incompatibility specificity only becomes effective at the time of the opening of the flower for it was shown that the pollen of unopened flowers of either type would germinate, develop, and effect fertilization in opened or unopened flowers of both types. Pollen of opened flowers, on the contrary, would fertilize unopened flowers only in those combinations bringing together pollen and stylar tissues of dissimilar incompatibility genotypes. It was thus demonstrated that the inhibitory constituents of the pollen, elaborated at anthesis, act as incompatibility-class-specific inhibitors of pollen germination and development. It was further shown that pollen of unopened flowers of both types contain quercetin but neither of the two inhibitory glycosides.

Studies of pollen germination *in vitro* provided definite proof of the inhibitory nature of specific pollen constituents. Pollen of unopened flowers of both incompatibility classes germinate readily in sugar solution (20 to 30 per cent) but germination of both types is completely suppressed by either rutin or quercetrin in a concentration of $1 : 10^6$. Pollen of opened flowers of neither type germinates in sugar solution, but pollen of each type germinates readily in an extract of the appropriate stylar tissue, i.e. *pin* pollen in extract of *thrum* styles and vice versa. The effectiveness of stylar extracts in removing the autoinhibition of pollen germination was in each case shown to result from the activity of a specific enzyme. The failure of germination of both types of pollen in sugar solution could also be overcome by treating the pollen with boric acid in a concentration of $1 : 10^4$. The efficacy of boric acid in pollen germination and in the negation of self-sterility due to incompatibility, an effect which had been known for many years (Schmucker, 1934) was shown in this case to result from the formation of inactive complexes of boric acid with the two inhibitory flavonol glycosides.

The occurrence of rutin and quercetin in the indicated tissues and filtrates of *Forsythia* was confirmed by supplemental tests with appropriate strains of *Chlamydomonas,* a green alga in which these compounds have been shown to have specific regulatory roles in the sexual process (Moewus, 1949; 1951).

The mechanism of pollen incompatibility specificity in *Forsythia* is schematically represented in Fig. 3.

A different type of terminal mechanism which accomplishes precisely the same end results as do the specific inhibitors of *Forsythia* has been described by Lewis (1943) in *Linum grandiflorum,* a species of the flax genus. In this species osmotic pressure characteristics of pollen and style are such as to exclude in incompatible combinations the physical conditions requisite to pollen germination and development. The osmotic pressures of styles and pollen, as measured by plasmolysis and tissue tension and expressed as "sugar percentage equivalents," are as follows: *thrum* style, 10 to 12; *pin* style, 20; *thrum* pollen, 80; *pin* pollen, 50. The difference in osmotic pressures of the tissues in both compatible pollen-style combinations is approximately 4 : 1, a differential which allows water uptake, germination, and normal development of the pollen. The differentials of osmotic pressure in the two incompatible combinations, however, are 7 : 1 in *thrum* $\times$ *thrum,* which permits pollen germination but causes the young pollen tubes to burst, and 2.5 : 1 in *pin* $\times$ *pin,* in which combination the pollen grains do not even swell upon the stigmatic surface. In the latter case osmotic pressure difference can not explain the total effect and it was suggested that differences in protoplasm colloids were probably responsible for the lack of imbibition of water by the pollen.

Yet another type of controlling mechanism is known to operate in certain other species having heteromorphic incompatibility and generally in those with homomorphic incompatibility, i.e. multiple-allele incompatibility. This type of control of specificity was postulated by East (1929) to be similar to immunity reactions in animals. Thus the product of a specific incompatibility allele in the pollen might act as an antigen to induce, in styles carrying the same allele, the production of an antibody which inhibits the growth of the antigen producing pollen but which has no effect upon pollen carrying any other allele. Lewis (1952) has recently demonstrated, in the pollen of *Oenothera organensis,* incompatibility specific substances which act as antigens in serological tests. Corresponding antibodies, reasonably assumed to be present in stylar tissues, have not yet been demonstrated. Lack of enhanced inhibition in successive pollinations indicated preformed stylar antibodies rather than their production in response to antigens carried by the pollen. That each incompatibility allele consists of at least two distinct subunits has been indicated both in studies with interspecific hybrids between self-incompatible and self-compatible species and in mutational studies (Lewis,

[129]

1949). In the latter case it was shown, in pollen subjected to x-rays prior to pollination in an incompatible combination, that the pollen specificity of a given allele could be altered or lost without any corresponding change in the stylar specificity of the mutated allele as expressed in a subsequent generation. The absolute selection in this work of altered pollen characteristics possibly accounted for the failure to find alleles with altered stylar specificities.

IV. INCOMPATIBILITY SPECIFICITIES IN THE HIGHER FUNGI

Multiple-allele incompatibility, as pointed out earlier, has evolved in only two groups of plants, the Angiosperms of the seed plants and the Basidiomycetes of the fungi. In this latter group it occurs only in the Hymenomycetes and Gasteromycetes, which groups comprise the "mushrooms," "bracket fungi," etc. and the "puffballs," "stinkhorns," etc. respectively. In the Hymenomycetes one or the other of two types of multiple-allelic incompatibility occurs in a large majority of all species: one-locus incompatibility, *bipolar sexuality,* and two-locus incompatibility, *tetrapolar sexuality* or *tetrapolarity.* Although the role of incompatibility factors in determining fertile mating specificity in tetrapolar species was first established by Kniep in 1920, recent and continuing work has revealed several additional mycelial interactions of high specificity which are directly attributable to the incompatibility factors. The ultimate mechanism which determines these various interactions is unknown; the correlations which are possible here between the genetics of incompatibility control and its ultimate effects, however, make tetrapolarity one of the more intriguing patterns of sexuality known among plants.

The essential characteristics of tetrapolarity result from (a) the independent assortment of incompatibility factors at two loci with the consequent production of four types of haploid progeny at meiosis and (b) the occurrence of an extended series of alternate and equivalent factors, i.e. multiple alleles, at each of the two incompatibility loci. The small-fruited wood-rotting mushroom *Schizophyllum commune* has probably been more intensively studied than any other and epitomizes the pattern of tetrapolarity.

The dikaryotic-diploid phase, the sporocarp or fruiting body, is characterized by two specific incompatibility factors at each of two loci, e.g. $A^1A^2B^1B^2$. Independent assortment of these factors at meiosis and the process of exogenous spore production yield haploid spores of four incompatibility factor constitutions in equal frequency:

$$A^1A^2B^1B^2 \longrightarrow A^1B^1,\ A^1B^2,\ A^2B^1,\ A^2B^2$$

In matings between haploid mycelia (homokaryons) derived from spores of these types the following reactions obtain:

(a) Unlike factors at both loci, e.g. $A^1B^1 \times A^2B^2$, allow mating with the reestablishment of the dikaryotic phase and the eventual completion of the sexual cycle.

(b) Common A-factors and dissimilar B-factors, e.g. $A^1B^1 \times A^1B^2$, yield reciprocally constituted heterokaryons throughout both mated mycelia.

(c) Dissimilar A-factors and common B-factors, e.g. $A^1B^1 \times A^2B^1$, yield a restricted heterokaryon in the immediate region of mycelial intermingling.

(d) Common factors at both loci, e.g. $A^1B^1 \times A^1B^1$, permit an interaction, inadequately characterized as yet, only under circumstances which preferentially favor the product of the interaction (*vide infra*).

These mycelial interactions, each to be discussed later in some detail, regularly occur among the progeny of each dikaryotic-diploid *stock* having a specific combination of four incompatibility factors. Each fruit collected from nature, however, contains its own specific combination of incompatibility factors; the duplication of single specific factors at either locus in different wild fruits is of the order of only 1 to 2 per cent and matings between progeny of different wild fruits accordingly yield dikaryons in practically all cases. The extent of the series of alternate factors at either locus in natural populations is not precisely known, but the best estimates indicate a minimum of the order of 100 at each locus (Whitehouse, 1949). The individual incompatibility factors of each multiple series may be considered as essentially equivalent as regards their determination of compatibility although very minor differences in reaction have been demonstrated between A-factors in a single species (Quintanilha, 1939).

The occurrence of a multiple-factor-series at each locus was recognized by Kniep (1922) shortly after his original description of tetrapolarity. Later he demonstrated that occasional monosporous mycelia carried incompatibility factors dissimilar to those of the dikaryotic-diploid parent in frequencies of approximately 2 per cent at the A-locus and 0.5 per cent at the B-locus (Kniep, 1923). These new factors were interpreted by Kniep as mutations at the incompatibility loci, a view which has since been generally accepted. Recent work, however, indicates a different origin of the "new" factors and provides some significant information concerning the genetic basis of the multiplicity of equivalent incompatibility factors at a single locus. The results of this work warrant consideration both for the direct bearing which they have on factor specificity and for the preliminary "exploration" of the incompatibility factor which they make possible.

A number of details given by Kniep concerning the origin of the "new" incompatibility factors led Papazian (1951) to question the validity of the commonly held interpretation of factor mutation. By tetrad analysis he was able to show in *S. commune* that the alleged mutations at the *A*-locus originated in the meiotic process and that they occurred in pairs of two new dissimilar factors along with the two original factors in single tetrads. Cross-mating of mycelia carrying the four *A*-factors of such a tetrad revealed complete interfertility between the different factors. Furthermore cross-mating between the two "new" factors of the tetrad yielded occasional tetrads in which the two original factors were now present as "new" factors. Representing factors differing from the two originals in the lower case and ignoring the *B*-factors involved, these results may be summarized as follows:

$$A^1- \times A^2- \xrightarrow{\text{meiosis}} A^1-,\ A^2-,\ a^1-,\ a^2-$$

$$a^1- \times a^2- \xrightarrow{\text{meiosis}} a^1-,\ a^2-,\ A^1-,\ A^2-$$

It was thus shown beyond reasonable doubt that the incompatibility factor is a compound genetic structure which functions as a single physiological unit and that "new" incompatibility factors originate not by mutation but by recombination of genetic units.

A rough approximation of the number of subunits comprising the incompatibility factor is possible from data included in Papazian's dissertation (1950) but which unfortunately were omitted from his published account (1951). In a total of about 300 tetrad and random progeny of a single stock, a total of eight recombined factors representing six distinct and interfertile classes were recovered. Thus a minimum of eight recombination classes, including the two classes represented by the two original factors, are involved and could be accounted for by three loci except for the vanishingly small probability of recovering both double cross-over classes in a sample of this size; four loci, which would require no double cross-overs here, appear to be the minimal number consistent with these recombination data. Calculations based upon the distribution of recombinants of the different classes and upon certain simplifying assumptions indicate four to ten loci within 5 per cent limits of probability (Raper, 1953). These findings, though inadequate to define the A-factor in precise genetical terms, indicate a sufficiently extended series of distinct loci to predict the total number of A-factors in a natural population without necessary recourse to multiple alleles at any single locus. A minimum of seven loci (128 possible recombination classes) would be required to provide the estimated 100 natural factors if no true multiple allelic series were involved.

The demonstration of the compound nature of the incompatibility factor reveals an interesting sort of specificity—a specificity which is dependent upon a genic combination in which, however, no single gene is required to determine specificity. This may be illustrated by the minimal four-locus factor although the same principle would apply equally to a more extended factor. The two original incompatibility factors and the recombination which yields "new" factors may be represented as follows:

$$
\begin{array}{ll}
 & \text{cross-over} \\
 & \text{at meiosis} \\
 & p^1\text{—}r^1\text{—}s^1\text{—}t^1 \quad \longrightarrow p^1\text{—}r^1\text{—}s^1\text{—}t^1 \ (A^1) \\
p^1\text{—}r^1\text{—}s^1\text{—}t^1 \ (A^1) \quad p^1\text{—}r^1 \quad s^1\text{—}t^1 \quad \longrightarrow p^1\text{—}r^1\text{—}s^2\text{—}t^2 \ (\, a^1) \\
\qquad\longrightarrow \qquad\qquad \times \\
 \quad p^2\text{—}r^2 \quad s^2\text{—}t^2 \quad \longrightarrow p^2\text{—}r^2\text{—}s^1\text{—}t^1 \ (\, a^2) \\
p^2\text{—}r^2\text{—}s^2\text{—}t^2 \ (A^2) \quad p^2\text{—}r^2\text{—}s^2\text{—}t^2 \quad \longrightarrow p^2\text{—}r^2\text{—}s^2\text{—}t^2 \ (A^2)
\end{array}
$$

Since parental and recombinant factors are interfertile in all combinations, the specificity of interaction can only be determined by the specificity of the entire combination, e.g. $a^1 \times A^1$ involves a duplication of p^1 and r^1 while $a^1 \times A^2$ involves duplication of s^2 and t^2. Furthermore the known extension of interfertility to include other recombinant classes as well can only be interpreted as the complete indifference of the system to the duplication of any specific component member or members and probably to the duplication of its members to any extent short of totality. Thus Papazian's conclusion, "a change in any allele in the A incompatibility group provides a functionally different incompatibility factor," based on a two-locus or three-locus factor and although probably correct for the revised factor, can not be accepted as proved for an incompatibility factor comprising more than four distinct loci. Further work will be necessary to establish the minimal genetic uniqueness required to determine the specificity of the entire compatibility factor. Whatever this minimum, however, the resulting factor-specificity rigorously determines specific behavior at the level of mycelial interactions.

At the mycelial level, it will be recalled, specificity of interaction is determined by factors at both incompatibility loci and the factors at the two loci may be related in four possible ways: (a) *compatible,* dissimilar factors at both loci, (b) *hemicompatible-a,* common factors at the *A*-locus, dissimilar factors at the *B*-locus, (c) *hemicompatible-b,* dissimilar *A*-factors, common *B*-factors, and (d) *noncompatible,* common factors at both loci. Specifically characterized heterokaryons, i.e. mycelia containing genetically dissimilar nuclei, are the products of compatible and both types of hemicompatible matings and probably of noncompatible matings as well. Brief description and comparison of these heterokaryons reveal a number of interesting and biologically significant phenomena.

Protoplasmic continuity is established in each type of mating by hyphal anastomoses, a phenomenon which is common and perhaps universal among higher fungi.

(a) *Compatible interaction* $(A^1B^1 + A^2B^2)$. Following the establishment of hyphal connections between two compatible homokaryotic mycelia, migration of nuclei from each mate into the other effects reciprocal fertilization. Once within the established mycelium, the invader nuclei rapidly migrate throughout the preexisting mycelium, dividing frequently en route, one daughter of each division remaining behind paired with a resident nucleus. As the wave of invader nuclei, proceeding at a rate several times that of vegetative growth (Buller, 1931), reaches the peripheral growing fringe of the mycelium, a pair of associated nuclei, the *dikaryon,* is established in each terminal rapidly growing cell. The dikaryon, constituted of one nucleus from each of the two mated mycelia, henceforth is maintained by the repeated simultaneous and adjacent division of its members. This process, *conjugate division,* fixes the ratio of nuclei of the two genetic types at 1 : 1 in the resulting dikaryotic mycelium, which commonly is composed of binucleate cells exclusively and upon which at each septum are evident the remnants of the apparatus of conjugate division, the clamp connection.

The dikaryotic mycelium is capable of indefinite vegetative growth and under continuing favorable conditions may produce successive crops of fruiting bodies during a period of years. Ordinarily only the dikaryotic mycelium produces fruiting bodies; the other types of heterokaryons are infertile.

One important characteristic of the dikaryon, of significance by comparison with the common-A heterokaryon to be described below, deserves mention at this point. Each nucleus of the dikaryotic pair masks by mutual complementation the specific genic inadequacies of the other so that dikaryons which contain no homozygous deficiencies look and react like wild type. For example the dikaryon resulting from the mating of compatible strains of *uracilless* and *niacinless* grows normally upon a minimal medium which contains neither uracil nor niacin as well as does a *wild* $\times$ *wild* dikaryon (Raper and San Antonio, 1954).

(b) *Hemicompatible-a interaction* (A^1B^1 : A^1B^2). In contrasts of mycelia having common factors at the A-locus, nuclear interchange and migration reciprocally convert each of the preexisting mycelia into a heterokaryon of a type which differs in several important respects from all other heterokaryotic types previously known among the higher fungi. The principle features of the common-A heterokaryon, which was originally described by Papazian in 1950 and which has been adequately characterized only during the past few months (Raper and San Antonio, 1954), are presented below.

Invading nuclei rapidly migrate into the established mycelium, heterokaryotizing it as the migration proceeds. The invading nuclei are restricted in their migration or division, however, so that they remain outnumbered by the original resident nuclei. Thus the heterokaryon originating from a single mate of a contrast contains a *predominant* (resident) *nuclear type* and a *secondary* (invader) *nuclear type*. The ratio of predominant: secondary nuclei ranges widely in different heterokaryons and in a given heterokaryon with age; ordinarily the ratio lies between 20 : 1 and 100 : 1, but occasionally may exceed 1000 : 1. Hyphal tips comprising 20-50 cells contain only predominant nuclei in most cases; both nuclear types occur in only about 5 per cent of the tips. The morphological features of the entire heterokaryon are determined by the predominant nuclear type.

In a single mating of two hemicompatible homokaryons, each having its own genetic peculiarities, there are accordingly established two specific heterokaryons, the nuclear structures of which bear a rough inverse relationship to each other. These two reciprocally constituted heterokaryons frequently are quite dissimilar morphologically as their growth characteristics are specifically determined by nuclei of different genotypes. These relationships may be represented in the following way:

$$A^1B^1 \times A^1B^2 \to (A^1B^1 : A^1B^2) =$$
$$[A^1B^1(A^1B^2)] + [A^1B^2(A^1B^1)]$$

[135]

A second feature of the common-*A* heterokaryon is the complete lack of mutual complementation between its component elements. For example a hemicompatible heterokaryon constituted of *uracilless* × *niacinless* strains will not grow upon minimal medium although as was shown earlier the same deficiencies in dikaryotic association are completely masked by complementation.

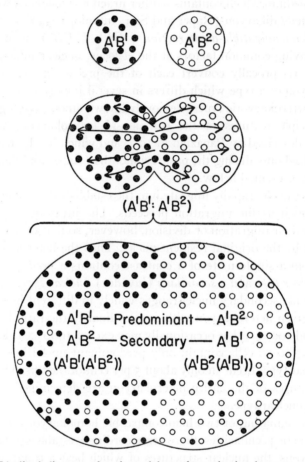

Fig. 4. Idealized diagram of reciprocal heterokaryotization between common-*A* homokaryons of *Schizophyllum commune* and the approximate distribution of nuclei of the two incompatibility types in the final heterokaryotic products of the interaction.

Lack of complementation moreover is only one expression of a specific interaction between the two components of the heterokaryon. Another and equally interesting expression may best be illustrated by outlining a simple experiment. A heterokaryon, constituted of two wild

[136]

strains, grows scarcely at all for a period of about a week upon a minimal medium in spite of the normal growth of both of its components upon the same medium. The addition to the minimal medium of some 28 specific vitaminlike substances enhances the growth of the heterokaryon to equal that upon a complete medium. An antagonistic interaction between the components of the common-A heterokaryon was thus shown to create for the nuclear association a dependence upon accessory supplementation which the component strains individually lack. That heterokaryosis seriously interferes with basic metabolic processes was further indicated by decreased growth and low sugar utilization of the heterokaryon as compared with homokaryotic and dikaryotic mycelia.

Morphologically the heterokaryon exhibits a depressed sparse habit with little or no aerial mycelium and is easily distinguishable under almost all conditions of culture from homokaryotic and dikaryotic mycelia. Microscopically it is distinguishable by the gnarled, knobby, irregularly branched habit of an appreciable but variable fraction of its hyphae.

Cultural stability is perhaps the most surprising of the several characteristics of the common-A heterokaryon for it can be subcultured indefinitely without the sectoring of its component strains as homokaryotic mycelia—in spite of the fact that growing elements which are demonstrably homokaryotic continually occupy peripheral and strategic positions for the escape from an association which by all reasonable criteria is acutely disadvantageous.

Of perhaps parenthetical interest in the present discussion is the occurrence in aging common-A heterokaryons of expressed single-gene mutations in frequencies ranging between 100 and 100,000 times the maximal spontaneous rate in homokaryotic mycelia. Present evidence indicates an automutagenic process which operates through the normal production of a filterable mutagenic agent(s) which preferentially effects a restricted number of loci in a regular and predictable manner (Raper and San Antonio, 1952).

(c) *Hemicompatible-b interaction* (B^1A^1 : B^1A^2). A limited heterokaryon confined to the immediate vicinity of the line of intermingling of the two mated mycelia has been described by Papazian (1950) in matings between strains having common factors at the B-locus. Heterokaryosis has not been rigorously proved in this interaction, but the many similarities between its mycelial product and that of common-A matings make it appear most likely that true heterokaryosis occurs. The preliminary cultural and mating studies of this presumed heterokaryon revealed two important differences between it and the common-A

heterokaryon described above: (a) the restriction of invasion of nuclei from each mate to a depth of the order of one millimeter or less into the mycelium of the other, and (b) the pronounced tendency of the heterokaryon upon subculture to sector quickly into its component homokaryotic strains.

(d) *Noncompatible interaction* (A^1B^1) (A^1B^1). Matings of biochemically wild strains of the same mating type, i.e. having common factors at both A-loci and B-loci, have to date provided no evidence whatsoever of interaction of any kind. Necessity, however, even in the affairs of fungi, provides a powerful incentive to inventiveness, for mixed inocula of *uracilless* × *niacinless* strains transferred to minimal medium produce a beautifully growing mycelium (Raper and San Antonio, 1954). Preliminary attempts to demonstrate heterokaryosis here were unsuccessful and growth may depend upon a stabilized and interacting mixture of pure homokaryons, each obtaining its specific requirement by diffusion from the other. But whether the association here is heterokaryosis or mixture, the fact of complete mutual complementation is of significance in the specificity of interactions of incompatibility factors.

Thus four distinct types of mycelial interactions result from the four possible combinations of incompatibility factors and constitute a more complex pattern than an all-or-none regulation of dikaryotization. Specifically different activities for the A and B factors are implicit in the determination of distinct reactions by the two hemicompatible combinations. The hemicompatible reactions thus appear to represent in frustrated isolation the separately inadequate contributions of the two incompatibility factors toward the establishment of the dikaryon and the completion of the sexual cycle. An attempt to delineate the mechanism of incompatibility control in tetrapolarity, however, is untimely. Detailed characterizations of the common-B heterokaryon and the noncompatible interaction, as yet unavailable, are essential parts of the puzzle.

V. CONCLUSIONS

Specificity of mating interactions in plants is thus seen to result from numerous genetic devices which characteristically impose restrictive blocks at various levels of organization and at various stages of developmental progression. The few examples cited here illustrate the range of specificity involvement in reproductive processes. Specificity of the compound incompatibility factors in *Schizophyllum* is determined by interactions between several individual genes while specificity of mating behavior in the same species is imposed by interactions between four distinct groups of genes, the incompatibility factors. The further elucidation of the mechanisms of these two specificities will be sought in genetical and physiological studies respectively. Interaction specificity in *Achlya* and related water molds has been studied so far at the level of the physiological aspects of the dual-purpose, differentiative-restrictive, hormonal mechanism with the expectation of the ultimate extension of the study to the level of biochemical definition. Specificities of pollen reactions can be defined in a few cases in terms of relatively simple terminal systems which may be termed either biochemical, as in *Forsythia,* in which specific inhibitors are known to be operative, or biophysical, as in *Linum,* in which osmotic differentials are determining.

A full appreciation of the over-all problem of specificity in the sexuality of plants can result only from the accumulation of more precise information. Specificities of critical importance, of many different kinds, and at every level of organization must occur in every plant with a sexual cycle and many new expressions of sexual specificity will undoubtedly be revealed before a simplifying integration is possible.

BIBLIOGRAPHY

Bateman, A. J. 1947. Number of S-alleles in a population. *Nature 160,* 337.
Buller, A. H. R. 1931. *Researches on Fungi. IV.* Longmans, Green & Co., London.
East, E. M. 1929. Self-sterility. *Bib. Gen. 5,* 331-368.
Kniep, H. 1920. Über morphologische und physiologische Geschlechtsdifferenzierung. *Verh. phys.-med. Ges. Würzburg 46,* 1-18.
Kniep, H. 1922. Über Geschlechtsbestimmung und Reduktionsteilung. *Verh. phys.-med. Ges. Würzburg 47,* 1-28.
Kniep, H., 1923. Über erbliche Änderungen von Geschlechtsfaktoren bei Pilzen. *Zeits. indukt. Abst. Vererb. 31,* 170-183.
Lewis, D. 1943. The physiology of incompatibility in plants. II. *Linum grandiflorum. Ann. Bot. (N.S.) 7,* 115-122.
Lewis, D. 1949. Incompatibility in flowering plants. *Biol. Rev. 24,* 472-496.
Lewis, D. 1952. Serological reactions of pollen incompatibility substances. *Proc. Roy. Soc.* (London) *B 140,* 127-135.
Mather, K. 1944. Genetical control of incompatibility in Angiosperms and Fungi. *Nature 153,* 392-394.
Moewus, F. 1949. Zur biochemischen Genetik des Rutins. *Port. Acta Biol. (A) Goldschmidt Vol.,* 161-199.
Moewus, F. 1950. Zur Physiologie und Biochemie der Selbststerilität bei *Forsythia. Biol. Zentralbl. 69,* 181-197.
Moewus, F. 1951. Die Sexualstoffe von *Chlamydomonas eugametos. Erg. Enzymforschung 12,* 1-32.
Papazian, H. P. 1950. The genetics and physiology of the incompatibility

alleles and some related genes in *Schizophyllum commune*. Doctor's thesis, University of Chicago.

Papazian, H. P. 1950. Physiology of the incompatibility factors in *Schizophyllum commune*. *Bot. Gaz. 112*, 143-163.

Papazian, H. P. 1951. The incompatibility factors and a related gene in *Schizophyllum commune*. *Genetics 36*, 441-459.

Quintanilha, A. 1939. Étude génétique du phénomène de Buller. *Bol. Soc. broteriana Ser. 2, 14*, 17-46.

Raper, J. R. 1939. Sexual hormones in *Achlya*. I. Indicative evidence for a hormonal coordinating mechanism. *Am. J. Bot. 26*, 639-650.

Raper, J. R. 1940. Sexual hormones in *Achlya*. II. Distance reactions, conclusive evidence for a hormonal coordinating mechanism. *Am. J. Bot. 27*, 162-173.

Raper, J. R. 1942. Sexual hormones in *Achlya*. III. Hormone A and the initial male reaction. *Am. J. Bot. 29*, 159-166.

Raper, J. R. 1942. Sexual hormones in *Achlya*. V. Hormone A', a male-secreted augmenter or activator of hormone A. *Proc. Nat. Acad. Sci. 28*, 509-516.

Raper, J. R. 1950. Sexual hormones in *Achlya*. VI. The hormones of the A-complex. *Proc. Nat. Acad. Sci. 36*, 524-533.

Raper, J. R. 1950. Sexual hormones in *Achlya*. VII. The hormonal mechanism in homothallic species. *Bot. Gaz. 112*, 1-24.

Raper, J. R. 1951. Sexual hormones in Achlya. *Am. Sci. 39*, 110-120.

Raper, J. R. 1953. Tetrapolar sexuality. *Quart. Rev. Biol.*, 28 : 223-259.

Raper, J. R. 1954. Life cycles, sexuality, and sexual mechanisms in fungi. *AAAS Symposium "Sex in Microorganisms,"* 42-81.

Raper, J. R., and J. P. San Antonio. 1952. Heterokaryotic mutagenesis in the tetrapolar fungus, *Schizophyllum. Rec. Gen. Soc. Am.*, p. 60.

Raper, J. R., and J. P. San Antonio. 1954. Heterokaryotic mutagenesis in Hymenomycetes. I. Heterokaryosis in *Schizophyllum commune. Am. J. Bot.*, 41 : 69-86.

Schmucker, T. 1934. Über den Einfluss von Borsäure auf Pflanzen, inbesondere keimende Pollenkörner. *Planta 23*, 264-283.

Whitehouse, H. L. K. 1949. Multiple-allelomorph heterothallism in the fungi. *New Phytol. 48*, 212-244.

Whitehouse, H. L. K. 1950. Multiple-allelomorph incompatibility of pollen and style in the evolution of the Angiosperms. *Ann. Bot. (N.S.) 14*, 119-216.

VII. GENERAL ASPECTS
OF IMMUNOLOGICAL REACTIONS
WITH BACTERIA AND PROTOZOA

BY JAMES A. HARRISON[1]

THIS presentation is limited to considerations of reactions between microorganisms and specifically related antibodies as these agencies are mixed *in vitro*. Excluded from direct discussion are the very important relationships between microorganisms and antibodies as they come into contact within the bodies of animal hosts.

The immunological reactions of agglutination, precipitation, and complement fixation have been very widely applied for more than a half-century in clinical laboratories as methods for the determination of the identity of infecting organisms. They have also been the favored reactions of researchers in the study of a wide variety of microorganisms. By comparison other direct actions of antibodies upon microorganisms, with the exception of the opsonocytophagic and the Quellung reactions, have been rarely employed in clinical and research laboratories. These other direct reactions between microorganism and antibody are nonetheless readily observed and in several instances have proved to be the most suitable reaction for application to some problems. They include (1) the immobilization of motile organisms, (2) the accumulation of semisolid product about the periphery of individual cells, and (3) a rather sharp interruption in the normal process of growth by division.

I. THE IMMOBILIZATION REACTION

Loss of motility upon exposure to homologous antiserum was observed in studies of motile forms of enteric bacteria by Gruber (1896), of paramecia by Roessle (1905) and Masugi (1926), and of other species of ciliated free-living protozoa by various workers including Schuckmann (1920) and Robertson (1934, 1939a). An effect superficially similar to the paralytic action is also produced on a variety of motile organisms by fresh serum collected from nonimmunized experimental animals, but the reactions are not the same. Immobilization caused by serum from nonimmune animals and humans is dependent upon the

---

[1] Department of Biology, Temple University. Appreciation is expressed to Elizabeth H. Fowler who has assisted with the new work at Temple University reported in this chapter.

[141]

presence of the nonspecific plasma component, complement, if it is not indeed due to its direct action, and it is destroyed by the removal of complement. On the other hand it was demonstrated by Masugi (1926) that the immobilizing effect of homologous immune serum is highly paralytic for paramecia even though all of the complement normally present has been destroyed. Further the immobilizing effect of serum is consistently a character in complement-free preparations from immune animals, while the paralytic effect of complement-bearing nonimmune serum is variable among individual serum donors (Harrison, Sano, Fowler, et al., 1948).

The paralytic effect of immune serum on enteric bacteria was observed by Gruber (1896). Smith & Reagh (1903) suggested that the paralysis of flagellated forms of salmonella was due to an entangling of flagella. They noted that agglutinates formed in antiserum prepared against a nonflagellated variant of the organism retained their motility. Pijper (1941) has in more recent years illustrated this paralytic action very strikingly in dark field microcinematographs of typhoid bacilli suspended in antibody for flagellated strains of the species.

Special events leading to a loss of motility are in part readily observed in several species of free-living ciliates upon exposure to complement-free homologous immune serum. The initial change noticeable in the paramecia, the colpoda, and the tetrahymena is the development of a small globular accumulation of semisolid material at the distal ends of one or more cilia; as time goes on a larger number of cilia show these accumulations and the accumulations increase in volume. Random contact between the terminal ends of cilia which show these structures very commonly results in the ends becoming stuck together, and one can see with careful observation that the afflicted animal is making an heroic effort to disjoin them. With a suitable preparation under the phase contrast microscope it can be seen that two or more cilia have formed loops which resemble a furculate, a tripod, or a multipod structure (see Plate I). Movement of the afflicted animal is for a time quite erratic and may include rapid spins of short duration around the longitudinal axis without motion forward or backward or very short or sharp thrusts forward or backward without rapid spinning around the long axis. The graceful and smooth movement of these animals has been disturbed and in general has been quite markedly reduced in speed. Involvement of still additional cilia in this way leads to further reduction in speed until finally the animal has lost all power of movement and simply can not move no matter how vigorously he is stimulated. The over-all speed of the reaction is dependent to a large degree on the strength of the antibody involved: stronger antibody concentrations give quicker reactions of ciliary involvement and paralysis; weaker concentrations give slower reactions. The extent of the reaction is also dependent upon the strength of antibody concentration; complete paralysis may never occur in weaker dilutions of antisera and the accumulation of solid material at the ends of cilia may be very little.

The relative practicability of the agglutinative, precipitative, complement-fixing, and immobilizing action of immune sera for studies on antigenic organization of the paramecia was determined in our laboratory several years ago (Bernheimer and Harrison, 1940). It was found that the immobilization reaction presented these advantages: (1) the antibody had a higher paralytic titer than the complement-fixation and precipitative titers and was somewhat higher than the agglutinative titer; (2) the immobilization reaction gave highly consistent reactions in triplicate preparations, as did the complement-fixation and precipitative reaction, whereas the agglutinative reaction was quite inconsistent in degree if not in occurrence; (3) the immobilization reaction could be prepared with far less effort than was required for the precipitative and complement fixation tests; and (4) the immobilization reaction permitted one to determine immediately the reactivity of each individual in the test population; this would be quite difficult to determine with the other reactions, although it is certain that one could fish individual animals of this size to isolated cells and run complement fixation and the usual type of precipitative test on them.

The application of the immobilization reaction for the determination of antibody to the spirochete of syphillis, *Treponema pallida,* was reported by Nelson and Mayer (1949). The effect reported is different from those just reviewed with the ciliated protozoa in that with the spirochete the presence of whole complement apparently is essential to the reaction whereas whole complement is not necessary in protozoan reactions. The refinement of this technique and final proof of its immunologic relation to spirochetal infection may provide the clinical laboratory with an excellent tool for diagnosis.

The high degree of specificity of the immobilized reaction is well illustrated in the results of cross reactions between several strains of the three species of paramecia in the aurelia group, i.e. *Paramecium aurelia, Paramecium caudatum,* and *Paramecium multimicronucleatum* (Bernheimer and Harrison, 1940, and Harrison and Fowler, 1944). It is also well illustrated in the correlation of antigenic group and mating type found in the species *Paramecium aurelia* (Bernheimer and Harrison, 1941).

The specifiicity of the immobilization reaction has been the foundation for the separation of antigenic mutants which arise in populations of bacteria and protozoa. The isolation of the Beta variants from monophasic strains of species in the genus *Salmonella* is accomplished with the technique of Wassen (1930, 1935) by fishing individuals or groups of cells which swim away from a mass of nonmutants that have been paralyzed by appropriate antibody. A similar technique has been applied in isolation of variant individuals in clones of paramecia (Harrison and Fowler, 1945a).

II. EXUDATION OF THE TEST ORGANISMS

The semisolid material which was first noted as a small globular accumulation at the distal ends of one or more cilia in the immobilization of protozoa often aggregates in a quantity sufficient to form a thick envelope about the entire animal—this is particularly true with the tetrahymena and the colpoda but is less readily demonstrable with the paramecia. The product formed by the paramecia is very soft and slimy. It is formed about the paramecium very rapidly but usually is swept backward along the surface of the animal by ciliary action about as rapidly as it is formed. In individuals which are partially paralyzed it appears as a long trailing stream of slimy material which picks up bacteria and other particulate elements of small size which happen to be in the suspension. These particulate elements commonly slip backward along the trailing slime to form a bulbous enlargement at the end. The character of the trailing material, in general shape and action, is something like the sea anchor that is so familiar to natives in this general area.

The slimy accumulation is an antigen-antibody product. It has been found to appear in any and all combinations of strains of paramecia and antisera which give the familiar precipitative and complement fixation reactions but it has never been observed in many other situations which are disturbing for the individual animal. Further evidence that this is an antigen-antibody product shall be given later in a discussion of some observations on the bacteria. While the accumulation of this slimy material first is seen at the distal ends of the cilia of the reacting animal, it has not been determined that all of the antigen donated to the reaction by the paramecium is poured out through these structures—it is possible that some of it passes through other parts of the peripheral surfaces.

The spectacular character of this reaction is accompanied by a striking loss in the volume of the paramecium. Camera-lucida drawings of a large number of individual cells following exposure to antiserum have been made and compared with drawings of unexposed individuals—on some occasions the same individual cells have been drawn before and after the treatment. Some of the preparations have been fixed in Schaudinn's fluid and stained; others have been fixed but not stained; still others have been untreated except by antibody. The cross-sectional and longitudinal dimensions of the drawings were taken into consideration in estimating the volume of the cells. On the average the cross-sectional area of antibody treated animals has been one-fifth to one-fourth less than that of the unexposed control animals. Rough calculation of volume based on a formula for the volume of a cylindrical structure with rounded ends indicates that one third or more of the total volume of the animal may be lost in its reaction to antiserum.

A similar reaction is observed in mixtures of tetrahymena and colpoda in homologous antisera (see Plate III, Figs. 1 and 2). The accumulations on these animals again is seen first at the terminal ends of the cilia. The product formed about the colpoda is a firmer jelly than the product about the paramecia—the product about the tetrahymena is often firmer still and sometimes is hard and cretaceous. The formation of trailing streams of slime described for the paramecia has not been observed in these smaller ciliates. On the other hand the product formed remains close to the surface of the animal and usually forms an enveloping cover. The first account of this in the literature is apparently the report of Robertson (1939a). She noted the formation of shell-like structures about the test organisms upon exposure to antiserum and further noted that from time to time the entrapped animal could finally escape from his enclosure and swim away and leave behind a perfect mold of his periphery. The mold not only showed clearly the striae of the body surface but also the pores from which the animal had withdrawn the cilia could be seen clearly.

The escape of a tetrahymena from the shell is a dramatic event. Following the full development of the shell and after a period of delay in which the entrapped organism exhibits no movement the animal rouses to show infrequent pulses of very slight movement. As time goes on these pulses become more frequent and the movement becomes more extensive, but the movement is very sharply limited to short turns within the shell until finally the animal suddenly spins for a second or so in either direction on its longitudinal axis and darts out of the enclosure.

On one occasion the author ran a continuous observation over a 4-hour period on a set of twins which had become enveloped in the shell at the time of division of the mother cell. The period of the observation ran from the forty-sixth to the fiftieth hour after the original antigen-antibody reaction. At the beginning of the observation the twin cells were about equal in appearance and each was showing about the same small amount of movement. Again the activity of the entrapped cells began to increase until each was struggling rather severely from time to time to escape. One of these cells then began to show a diminished activity and finally became motionless. The other twin kept the movement up with increasing frequence and finally escaped to leave behind his intended sarcophagus attached to the tomb which contained his dead brother. The escaped fellow swam about quite as merrily as do other young cells of the same race who have had no such experience.

The accumulation of the enveloping jelly about the colpoda is somewhat more uniform in amount and more extensive than about the tetrahymena. Also escape from the product seems to be a little easier for the colpoda since it occurs after shorter periods of waiting and since on the whole the colpoda more commonly escape than do the tetrahymena. It should be noted that the antibody concentration required for this reaction is somewhat higher than the concentration required for immobilization and agglutination of the animals.

It should be emphasized that the tetrahymena and the colpoda regain full capacity for movement and reproduction after escaping from their reaction product and that the paramecium can sometimes be nursed back to vigorous health with proper care. It should also be emphasized that so long as these escaped animals and their progeny remain constantly in the antibody similar to that of the initial reaction they remain insensitive to its presence. However, it commonly occurs that shortly after these individuals are removed to culture fluid which is free of antibody they regain a sensitivity to it. This is especially true of the tetrahymena and the colpoda.

This fact suggests that the antibody either inhibits synthesis of the antigenic component of the cell that is responsible for the reaction or that it leaches the antigen from the cell as rapidly as it is formed and that the quantity available for accumulation to paralyzing amounts of reaction product is never adequate for the reaction.

The development of solid accumulations of antigen-antibody product in and about larvae of the tapeworm *Nippostrongylus muris* when found in the tissues of immune and hyperimmune rats was reported by Taliaferro and Sarles (1939). Similarly the development of a gelatinous accumulation about the tails of the larval form of the frog *Rana pipiens* was noted by Parkes (1946) when these forms were exposed to homologous antisera and about larval forms of *Schistosoma* by Vogel and Minning (1949). These reactions seem to be quite similar to those just described for the protozoa.

A search for evidence of similar reactions with the bacteria in our laboratory has been confronted with some difficulty. Attempts to show such reaction with cells of the genera *Salmonella* and *Bacillus* have been made over a period of several years. Accumulations of solid substance about the periphery of salmonella have been noted with regularity in some species and inconsistently with other species. However, the possibility that the accumulation of solid substance about these tiny organisms is not an antigen-antibody product has not been ruled out. An increase in the refractility of various species of these organisms upon exposure to strong homologous antibody has been noted as a regular occurrence when observed with the phase contrast microscope. The optical character of the surfaces of these cells is very similar to the character of the small globules or reaction product which are found at the ends of the cilia of paramecia which are bathed in antibody. It may be that a thin accumulation of antigen-antibody product over the entire surface of the bacterium is responsible for this change in refractility. More success in searching for the reaction has attended our studies with the bacilli. Spores of the species *Bacillus subtilis* have given excellent reactions (see Plate III, Figs. 3 and 4). However resistant these structures may be to the effects of high levels of heat, of strong acids, and alkalis, they are very sensitive to the effects of strong antibody.

Shortly after exposure to antibody there is found an accumulation of refractile material about the periphery of the spore. Such accumulations do not occur about these spores when exposed to antisera which do not specifically agglutinate the spores or bring about the formation of precipitates with water extracts of the organisms.

The cover slip slide preparations for these observations must be very thin. It has been best to place the material to be examined on a thin film of agar which covers about half the diameter of the cover slip to be used. The cover glass is sealed on the slide with petroleum jelly and is pressed firmly against the agar film. The area under the cover glass is then searched for spores which are immediately on the lower surface of the glass.

The amount of reaction product found about the spores is variable from individual to individual. In some instances it has approached relatively the amount of reaction product found about the colpoda. The reaction develops rapidly at room temperature incubation. Generally it develops to fullest extent within fifteen minutes and sometimes within five minutes after initial exposure to antibody.

The strength of antibody required to induce the reaction here as in the case of the similar reactions in the tetrahymena and colpoda is greater than the strength required for gross agglutination and here again complement is not needed for the reaction. One lot of antiserum which had a gross agglutinating titer of 1 to 40 would bring about the exudation phenomenon only in dilutions of 1 to 16 or less; another lot which had a gross agglutinating titer of 1 to 160 would bring about the phenomenon in dilutions of 1 to 32 or less.

The reactions with these spores, as with many individuals of the ciliated protozoa, fails to destroy the viability of the body. Even spores which have undergone very extensive reaction seem to germinate quite as rapidly as untreated spores, and the emerging vegetative cell grows and divides apparently without hindrance.

The nature of the antigenic component in the reaction product which accumulates about the spores has not been determined by direct chemical analysis; however, it is evident that it contains a large moiety of carbohydrate in some form. This opinion is based upon several considerations.

Spores which have been treated with heat applied at levels of 85 to 100°C. for periods of 10 or more minutes do not form visible accumulations of solid material about the periphery upon exposure to strong antibody as untreated spores do. Neither do spores which have been exposed to 5 per cent solutions of KOH for 7 hours or more at room temperature (about 23°C.). On the other hand the fluid fractions from these treatments form precipitates with homologous immune serum. These fluid fractions give a positive Molisch test and so may be considered to contain a protein-free polysaccharide, a mucopolysaccharide, or a mucoprotein. This observation is in keeping with recent findings of Wynne and co-workers (1953), who reported that cultures of *Vibrio metschnikovii* contain a carbohydrate haptene that may easily be removed by heat.

Another bit of evidence which points up the protein-carbohydrate character of the reaction product about the spore is the fact that the accumulated material stains pink in toluidin blue stain. This, according to McClung (1929) and Galigher (1934), is a specific reaction for mucin.

The reaction product found about spores and about the tetrahymena is more soluble in 0.85 per cent saline than in distilled water. Spores which have been washed through as many as 20 changes of distilled water still show the material about the periphery. It disappears, however, with two or three subsequent washings in saline and is not found in spores which have been washed several times in saline without prior washing in distilled water. A difference in the solubility of the product formed with the protozoa is also found. Shells collected after the tetrahymena which formed them have escaped dissolve as a rule after four or five changes in saline solution but remain unchanged through as many as fifteen washings in distilled water. The antigen in these shells, incidentally, was reported by Robertson (1939b) to be a mucoprotein.

While this reaction seems at present not to be very practical as a method for serological determination of the identity of bacterial populations, it has been proved to be very practical in studies of some of the protozoa and the metazoa and in at least one instance has proved to be the most suitable method for study. Reference is made to the demonstration of antigenic change which occurs in antigenically dissimilar individual paramecia during the process of conjugation. Several years ago Harrison and Fowler (1946) studied the phenomenon of conjugation in two antigenically dissimilar strains of *Paramecium bursaria*. These two strains were also dissimilar in content of the included zoochlorellae to the extent that one of the strains was deeply green in color while the other was colorless. It was found by exposure of the conjugants to appropriate antisera at various stages in the conjugation period that near the time for crossing of the pronuclei each individual conjugant picked up the antigenic character of its mate and the two cells became almost identical antigenically, i.e. so far as the surface reaction was concerned. The results of this study indicated that the process of conjugation involved a crossover of significant amounts of cytoplasm as well as an exchange of nuclear elements.

III. GROWTH DISTURBANCE

It has long been known that exposure to homologous antibody can bring about a sharp deviation from the normal growth pattern in bacteria. Charrin and Roger (1889) noted the formation of long threads of undivided cells upon the exposure of *Pseudomonas aeruginosa* to serum of an immune animal. Issaeff (1893) observed that long chains of pneumococci were far more prevalent in cultures grown in immune serum than in cultures grown in normal serum. He also noted that some of the cells in the chains were unusually enlarged. Pfaundler (1898) noted the formation of long threads of *Bacterium coli* and *Proteus vulgaris* upon exposure of these organisms to homologous antiserum. Pfaundler further stated that numerous indications of thread formation are to be found in the earlier literature of Gruber and Widal.

While there is no intention to deal with the *in vivo* serological reactions here, it should be pointed out that Taliaferro (1924, 1932) has shown that a remarkable interruption of normal growth pattern in broods of infecting trypanosomes occurs as the host develops immunity to the organism.

It is puzzling that sharp disturbances in the growth of paramecia and colpoda are rarely observed as a possible reaction to antibody exposure since this is a very striking occurrence in the somewhat closely related holotrichous ciliate *Tetrahymena* (Harrison and Fowler, 1945b). (See Plate II.)

Shortly after exposure of the tetrahymena to an appropriate concentration of antibody one finds that the number of cells in some very obvious stage of division is on the increase. This becomes very striking after three or four hours of incubation, at which time as many as 80 per cent of the individuals in the test sample are in a dystomic condition. In the earlier periods after exposure to antibody one finds pairs of cells which seemingly can not effect the separation at the conclusion of division that occurs in normal cultures. As time goes on an increasing number of unseparated pairs of cells is noted and there begin to appear some few chains of three or four cells smimming about with animation in the medium. The maximum number of undivided cells in chains observed by the author as a result of this reaction has been six.

What happens to these undivided pairs and chains? Eventually most of them successfully effect a separation from their brothers, but upon some occasions a pair or even a set in a chain will form into a binucleate or multinucleate individual which carries on still further some of the phenomena of growth. This leads to the development of multinucleate giant cells of variable morphology.

Most of the time these multinucleate giant cells maintain on their surfaces much of the morphology of the surface of a normal cell, i.e. the striation and ciliary arrangement of the surface bears a remarkable resemblance to that of individual cells. It looks more or less as though a number of cells have fused together and have at the same time retained individuality on the unfused outer surface. It has not been uncommon to find in this type of monster a bulging development of tissue which has to a striking degree the appearance of a normal cell; in fact the author has on many occasions watched such developments until they effect a complete septum from the remainder of the giant, pinch off from the mother cell, and swim away to become lost in other cells of very typical morphology.

Less frequently these giant cells grow into structures which show very little of the surface morphology of normal cells except ciliary structures, and these are not arranged in orderly rows. One of these giants was compressed between cover slip and slide without breaking until 16 individual nuclei and 16 contractile vacuoles were counted under the phase contrast microscope.

The mechanism by which the formation of pairs, chains, and giant cells comes about under the influence of homologous antibody in the tetrahymena is not clear. It is interesting to speculate that the antiserum may interfere with some element essential for the conclusion of division and separation. Perhaps there is an interference with the synthesis of a protein or a carbohydrate that is essential for the development of a complete septum. Perhaps there is some other interference which does not allow for the development of adequate amounts of cytoplasm within the pellicle of the organism. Perhaps it is due to a leaching of cytoplasmic component from the cell.

The leaching of cytoplasmic component from the cell would be the most attractive explanation of the dystomic reaction were it not for the fact that in the more or less closely related organisms colpoda and paramecia there is evidence of an even greater loss of cytoplasmic component and yet these organisms very rarely show dystomic forms in homologous antisera. Of course it is entirely possible that the surface structures of the tetrahymena are more rugged than those of the paramecia and the colpoda and so should better resist any breaks which would result in quick death of the organism. Were this true and if the dystomic reaction in tetrahymena is in fact due to the loss of too much cytoplasm to fill the pellicular structure, one should find some evidence of crenation of the pellicle. Now it is not difficult to find sharp departures from normal curvature in some of these undivided pairs and chains of the tetrahymena. However, it can not be said that this irregularity in the surface is due to an insufficient volume of cytoplasm to fill the pellicle. It may be due on the other hand to unequal aggregation of exteriorly displaced antigen-antibody product at different areas of the cells.

IV. A CONSIDERATION OF THE MECHANISM OF
SEROLOGIC AGGLUTINATION

All of these observations point to the fact that the ciliated protozoa and other forms of organisms, including some of the bacteria, show striking changes particularly associated with the surface of the organism upon exposure to antiserum. They also show clearly that these changes quite commonly include the accumulation on the surface of greater or lesser amounts of reaction product which is sticky in some degree.

In the early work on the paramecia in our laboratory the observation that the trailing slime which formed on these animals during exposure to antiserum would pick up a large number of bacteria normally present in the test material, and would pick up other small objects which might be added, compelled us to question the nature of the forces which bring about the attachments. Could it be that the entrapment of bacteria in the slime was due in part to the presence of antibacterial elements in the antiparamecium rabbit serum? We were never able to use bacteria-free preparations of paramecia for immunization although the animals were rid of a great deal of the bacterial population of fresh cultures by repeated washing in sterile fluid. It was reasonable to suppose then that the antiparamecia serum might contain antibacterial substance and that this was partly responsible for the bacterial entrapment in the slime. Several efforts were made to demonstrate antibacterial substance in the paramecia antisera. Each of several strains of bacteria which had been isolated from gross cultures of paramecia and from some individual animals which had been washed by the Parpart (1928) technique were examined for agglutinability in appropriate paramecia antisera. Most of the strains failed to agglutinate in any dilution of the serum and the few that clumped in the serum did so in the very low dilutions of 1 to 2 and 1 to 4.

Further evidence of the nonspecific nature of the entrapment of bacteria is the fact that the reaction product formed on the colpoda and the tetrahymena by antisera prepared with bacteria-free cultures of the animals would also pick up bacteria and other micellar materials when they were placed in the mixtures.

The nature of the cohesive forces which bring about the clumping of organisms upon exposure to antibody has been a problem of great interest since the reaction was first demonstrated by Gruber (1896). Gruber emphasized that the clumping of bacteria in antiserum was due to an increase in the stickiness of the surfaces. In fact the name, "Glabrifacine,"

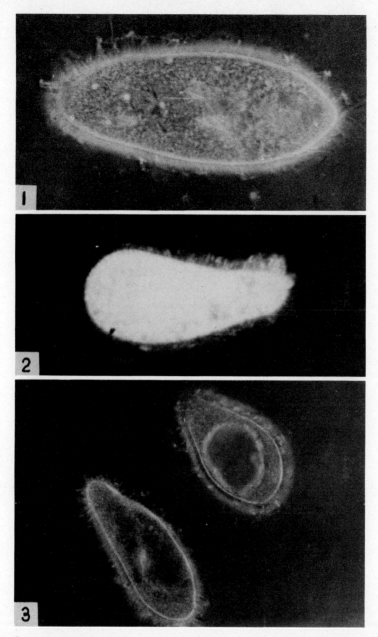

PLATE I.

Fig. 1. Unaltered phase contrast photomicrograph of *Paramecium aurelia* taken after short exposure to homologous antiserum. Note tripod or furculate arrangements of cilia to the left on the upper margin.

Fig. 2. Unaltered dark field photomicrograph of *Glaucoma pyriformis* after short exposure to homologous antiserum. Note matting and thickening of cilia, particularly toward distal ends along upper margin.

Fig. 3. Unaltered phase contrast photomicrograph of *Tetrahymena geleii* after exposure to homologous antiserum. Note enlargements near or at the ends of cilia.

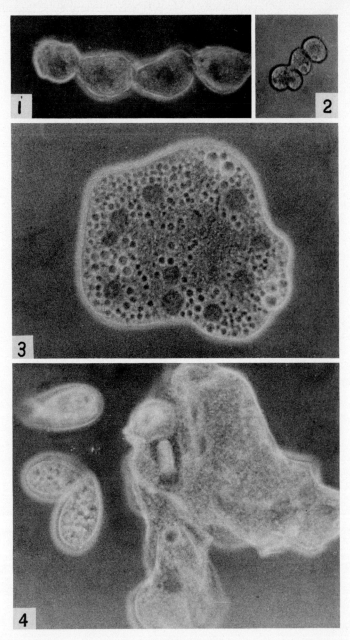

PLATE II. Unaltered phase contrast photomicrographs of dystomic forms developed in *Tetrahymena geleii* during exposure to homologous antiserum.

Fig. 1. Note irregularity and possible crenation in pellicular surfaces. Also note three globular enlargements at distal ends of cilia on upper margin of the second from the left part of the chain.

Fig. 2. Irregularity in development of chains. Forms such as this often grow into giant cells.

Fig. 3. Optical plane through flattened giant cell. Note ten globular nuclei (sixteen nuclei were found in microscopic examination of this giant).

Fig. 4. Three cells of normal shape and size after recovery from action of antiserum, and one giant cell.

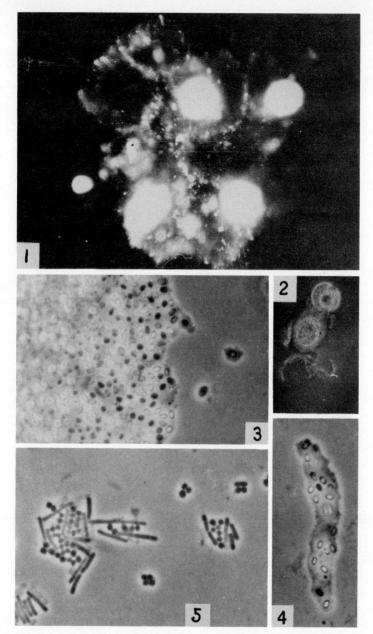

PLATE III.

Fig. 1. Dark field photomicrograph of a clump of *Glaucoma pyriformis* cells formed upon exposure to homologous antiserum. Note large amount of jellylike material about each of the four cells. Note pockets in the jelly mass from which recovered individuals have escaped.

Fig. 2. Phase contrast photomicrograph of *Tetrahymena geleii* after exposure to homologous antiserum. Note shell about the lower individual of the pair of cells and note two empty cells from which recovered individuals have escaped.

Figs. 3 and 4. Phase contrast photomicrographs of spore of *Bacillus subtilis* after exposure to homologous antiserum. Note accumulation of solid material between and around spores. Note also the distance betweeen individuals in the agglutinated masses.

Fig. 5. Phase contrast photomicrograph of mixed agglutination. Clumps formed in mixtures of *Salmonella paratyphi A* and *Staphylococcus aureus* during exposure to a mixture of homologous antisera. Note presence of both the rod and spherical organisms in the clumps.

which he gave to the antibody was defined parenthetically as "klebrig-macher." Bordet (1920) suggested that the aggregation of clumps was due to a reduction in certain physiochemical forces at the surface of the bacteria which tend to make them repel one another and remain dispersed in the suspending fluid. This earlier concept of Bordet's has been more or less replaced as a favored explanation of serological agglutination by a theory of alternate binding between bivalent antibody and multivalent antigen to form cubic lattice net works. Heidelberger and Kendall (1935) and Pauling (1940) have presented persuasive evidence in support of this, the Marrack (1934) hypothesis.

A solution to the question of the nature of the reaction which leads to serological agglutination has been sought in several studies of mixed populations of living cells and their respective antisera. Mixtures of pneumococci and enteric bacilli were studied by Topley, Wilson, and Duncan (1935); mixtures of human erythrocytes and Friedlander's bacilli were studied by Abramson (1935); mixtures of human and chicken erythrocytes were studied by Hooker and Boyd (1937); mixtures of A and B groups of human erythrocytes were studied by Abramson, Boyd, Hooker, Porter, and Purnell (1945). Mixtures of several types of bacterial cells were studied by Harrison and Fowler (1952). The results of these studies have been at variance. Boyd (1947) suggests that dilute antibody concentrations tend to give homogeneous clumps while strong concentrations tend to give mixed clumps. Our own experience has been at some variance: sometimes the clumps found are quite consistently homogeneous; more commonly, however, the clumps are mixtures of both types of the test organisms. We have not found less heterogeneity in clumps formed in high dilutions of antibody than in low dilutions. (See Plate III.)

One additional bit of evidence throws some light on this question. We have found that the distance between individual spores in serological agglutinates is fairly uniform from clump to clump and from spore to spore within the clumps. This distance is from three-quarters to the full diameter of the spore. The spatial proximity seems not to be influenced by the dilution of antibody used; those clumps formed in the higher dilutions which give incomplete gross agglutination have about the same spatial arrangement as those formed in strong concentrations of antibody. This observation can not readily be explained with the Marrack hypothesis unless one assumes that antigen leached from the spores forms with antibody a long bridgework spanning between the cells or unless the position is taken that the antibody is very long. The Bordet hypothesis

[153]

seems even less likely as an explanation of the observation. On the other hand a speculation that the spores are held together in clumps by the stickiness of the antigen-antibody product which surrounds them is an appealing idea. My personal feeling is that serological agglutination of microorganisms is a matter of nonspecific stickiness which in turn is due to a specific combination of antigen and antibody.

The Marrack and Bordet hypotheses are particularly appealing in that they provide a very logical explanation of the prozone phenomenon of serological agglutination. The Gruber hypothesis on the other hand provides no adequate explanation of the prozone unless it is to be found in the degree of stickiness of the antigen-antibody product formed at different levels of antibody concentration. One fascinating possibility is that the antigen-antibody product formed in strong concentrations of antibody may be more cretaceous and less sticky than that formed in weaker dilutions. This in fact does seem to be the case in the products formed about the tetrahymena. However, there is no evidence that this is true of the product formed with other microorganisms.

BIBLIOGRAPHY

Abramson, H. A. 1935. Mixed agglutination. *Nature 135,* 995.

Abramson, H. A., W. C. Boyd, S. B. Hooker, P. M. Porter, and M. A. Purnell. 1945. The specificity of the second stage of bacterial agglutination and hemagglutination. *J. Bact. 50,* 15-22.

Bernheimer, A. W., and J. A. Harrison. 1940. Antigen-antibody reactions in paramecium: the aurelia group. *J. Immun. 39,* 73-83.

Bernheimer, A. W., and J. A. Harrison. 1941. Antigenic differentiation among strains of *Paramecium aurelia. J. Immun. 41,* 201-208.

Bordet, J. 1920. *Traité de l'immunité.* Masson, Paris.

Boyd, W. C. 1947. *Fundamentals of Immunology.* 2nd Ed. Interscience Publishers, Inc., New York.

Charrin, A., and C. Roger. 1889. Note sur le développement des microbes pathogènes dans le sérum des animaux vaccinés. *Comp. Rend. Soc. Biol. 2,* 667-669.

Galigher, A. E. 1934. *The Essentials of Practical Microtechnique.* Albert E. Galigher, Inc., Berkeley, Calif.

Gruber, M. 1896. Theorie der activen und passiven immunitat gegen cholera, typhus und verwandte krankheitprocesse. *Munchen Med. Wochenschr. 43,* 206-207.

Harrison, J. A., and E. H. Fowler. 1944. Serologic differentiation of the three species of paramecium in the aurelia group. *Anat. Rec. 89,* 20.

Harrison, J. A., and E. H. Fowler. 1945a. Antigenic variation in clones of *Paramecium aurelia. J. Immun. 50,* 115-125.

Harrison, J. A., and E. H. Fowler. 1945b. An antigen-antibody reaction with Tetrahymena which results in dystomy. *Science 102,* 65-66.

Harrison, J. A., and E. H. Fowler. 1946. A serologic study of conjugation in *Paramecium bursaria. J. Exp. Zool. 101,* 425-444.

Harrison, J. A., M. E. Sano, E. H. Fowler, R. H. Shellhamer, and C. A. Bocher. 1948. Toxicity for paramecia of sera from cancerous and non-cancerous persons. *Fed. Proc. 7,* 306.

Harrison, J. A., and E. H. Fowler. 1952. Serological agglutination of mixed populations of bacteria. *Bact. Proc.,* page 95.

Heidelberger, M., and F. E. Kendall. 1935. The precipitin reaction between type III pneumococcus polysaccharide and homologous antibody. *J. Exp. Med. 61,* 563-591.

Hooker, S. B., and W. C. Boyd. 1937. The nonspecificity of the flocculative phase of serologic aggregation. *J. Immun. 33,* 337-353.

Issaeff, B. 1893. Contribution à l'étude de immunité acquisé contre le Pneumocoque. *Ann. Inst. Past. 7,* 260-269.

Marrack, J. R. 1934. *The Chemistry of Antigens and Antibodies.* 1st Ed. Med. Res. Council, London.

Masugi, M. 1926. Über die wirkung das normal sowie das spezifisschen immunsieums auf die paramäzien. Über die immunitäte derselben gegen die beiden serumwirkungen. *Krankheitsforschung 5,* 375-402.

McClung, C. E. 1929. *Handbook of Microscopical Technique.* Hoeber, New York.

Nelson, R. A., Jr., and M. A. Mayer. 1949. Immobilization of *Treponema pallidum in vitro* by antibody produced in syphilitic infection. *J. Exp. Med. 89,* 369-393.

Parkes, A. S. 1946. Anti-tadpole serum. *Nature 157,* 164.

Parpart, A. K. 1928. The bacteriological sterilization of paramecium. *Biol. Bull. 55,* 113-120.

Pauling, L. 1940. A theory of structure and process of formation of antibodies. *J. Am. Chem. Soc. 62,* 2643-2657.

Pfaundler, M. 1898. Eine neue form der serum reaktion auf coli und proteus bacilliosen. *Cent. f. Bakt., Parasit. und Infekt'kr. Abt. I. 23,* 71-79.

Pijper, A. 1941. Microcinematography of the agglutination of typhoid bacilli. *J. Bact. 42,* 395-409.

Robertson, M. 1934. An *in vitro* study of the action of immune bodies called forth in the blood of rabbits by the injection of the flagellate protozoon, *Bodo caudatus. J. Path. Bact. 38,* 363-390.

Robertson, M. 1939a. A study of the reactions *in vitro* of certain ciliates belonging to the Glaucoma-Colpidium group to antibodies in the sera of rabbits immunized therewith. *J. Path. Bact. 48,* 305-322.

Robertson, M. 1939b. An analysis of some of the antigenic properties of certain ciliates belonging to the Glaucoma-Colpidium group as shown in their response to immune serum. *J. Path. Bact. 48,* 323-338.

Roessle, R. 1905. Spezifissche sera gegen infusorien. *Arch. f. Hygiene 54,* 1-31.

Schuckmann, W. 1920. Serologische untersuchungen an Kulturamöben. *Berliner Klin. Wochschr. 57,* 545-547.

Smith, T., and A. L. Reagh. 1903. The non-identity of agglutinins acting upon the flagella and upon the body of bacteria. *J. Med. Res. 10,* 89-100.

Taliaferro, W. H. 1924. A reaction product in infections with *Trypanosoma lewisi* which inhibits the reproduction of the trypanosomes. *J. Exp. Med. 39,* 171-190.

Taliaferro, W. H. 1932. Trypanocidal and reproduction-inhibiting antibodies to *Trypanosoma lewisi* in rats and rabbits. *Am. J. Hyg. 16,* 32-84.

Taliaferro, W. H., and M. P. Sarles. 1939. The cellular reactions in the skin, lungs and intestines of normal and immune rats after infection with *Nippostrongylus muris. J. Inf. Dis. 64,* 157-192.

Topley, W. W. C., J. Wilson, and H. T. Duncan. 1935. The mode of formation of aggregates in bacterial agglutination. *Brit. J. Exp. Path. 16,* 116-120.

Vogel, H., and W. Minning. 1949. Hüllenbildung bei Bilharzia-cercarien im serum bilharzia-infizierter tiere und menschen. *Zent. Bakt. (Orig.) 153,* 91-105.

Wassen, A. 1930. Sur une méthode d'enrichissement des bacillus paratyphiques, basée sur la mobilité et l'agglutination directe des bacilles dans le milieu. *Compt. Rend. Soc. Biol. 104,* 523-527.

Wassen, A. 1935. Essais d'application au *Vibron cholérique* de la méthode fondée sur la faculté de displacement des bactéries. *Bull. Mensuel de l'Office Internat. d'Hyg. Publ. 25,* 1-14.

Wynne, E. S., C. L. Gott, D. A. Mehl, and H. A. Norman. 1953. Serological studies with *Vibrio metschnikovii. J. Immun. 70,* 207-211.

VIII. SPECIFICITY
IN THE RELATONSHIP BETWEEN
HOST AND ANIMAL PARASITES

BY W. H. TALIAFERRO[1]

THE central problems in the specificity of the host-parasite relationship concern the mechanisms whereby each parasite is limited in its range of hosts and, in the range of susceptible hosts, is more or less restricted in its site of colonization and in its power to invade, reproduce, and survive. It is not surprising that such specific relationships exist. Once a parasite invades a specific host it becomes more or less isolated just as free-living species become geographically isolated when they are segregated on islands or behind other ecological barriers (cf. Baer, 1951). The effective isolation of a parasite generally increases as parasitism becomes obligatory and involves the entire life cycle of the parasite. In time evolutionary changes would be expected to produce specific adaptations to the particular host environment in which the parasite is segregated. In this sense the host becomes equivalent to the biotope of the free-living species. It is, however, a peculiar biotope in that it is a living organism in which, where parasitism is widespread, it similarly evolves specific responses and adaptations to invasion by the parasite.

In this chapter consideration of the mechanisms of the host-parasite relationship will be limited to endoparasites; also, those situations in which purely ecological or similar factors prevent contact between a parasite and a potential host will be excluded. Assuming that there is effective contact between parasite and potential host, the biochemical and physiological aspects of the host-parasite relationship fall into two main categories: (1) the factors of parasite invasiveness which permit the parasite to penetrate and maintain itself in the host, and (2) the factors of host immunity, both innate and acquired, which operate reciprocally to prevent invasion and colonization of the host. Although virulence is generally associated with parasite factors and immunity with the host, the reciprocal relationship of the two often makes impossible clear-cut differentiations. Thus certain pantothenic-acid-deficient hosts do not suffer from malaria as do normal hosts. Assuming that this is a direct result of the need for this vitamin by the parasite, the situation can equally well be considered an athreptic host immunity resulting from the absence

---

[1] Department of Microbiology, University of Chicago.

[157]

of a sufficient concentration of the vitamin or to a lack of parasite virulence due to its inability to synthesize the vitamin. Finally, extraneous factors may simulate a decrease in virulence of the parasite or an increase in the activity of innate immunity by the host. An example of this is the antibiosis which some investigators believe to occur between certain bacteria and fungi in the human intestine.

The interplay of these diverse factors results in all degrees of parasitism between a given parasite and a potential host. At one extreme there are unquestionably many cases of absolute innate immunity in which the parasite can not make the initial invasion of the potential host. These may arise from any number of physical and chemical barriers or from some structural or chemical deficiency of the parasite. At the other extreme there are a few cases, especially in the laboratory, in which the parasite probably can not only invade and reproduce at its maximum capacity but almost all of its progeny can survive. Such associations can not last long as they inevitably result in the death of the parasite as well as the host. In fact no host infected with any endoparasite or most ectoparasites can continue to survive in the absence of some mechanism limiting increase of the parasites.

In considering such a complex range of factors as those involved in the interplay of host and parasite the discussion will necessarily be limited to a few illustrative examples.

I. FACTORS PRIMARILY ASSOCIATED WITH HOST IMMUNITY

Host immunity is usually divided into innate immunity, which involves inborn constitutional factors, and acquired immunity, which results from immunizing procedures. The essential characteristic of acquired immunity is associated with host antibodies which react specifically with the antigen or antigens of the parasite. They may be acquired (a) actively by infection with a specific parasite (Fig. 3) or other organism having common antigens or by immunization with a vaccine or other material containing the proper antigens, or (b) passively by injecting the host with serum containing specific antibodies. Innate and acquired immunity are frequently difficult to differentiate and after acquired immunity supervenes, they act together.

Antibodies confer on acquired immunity the high specificity of antigen-antibody reactions and are the best known of the specific factors in the host-parasite relationship. In sharp contrast the mechanisms of innate immunity constitute an extremely complex agglomerate of mechanisms which are difficult to study and have been very poorly analyzed. The author has frequently referred to them as the sum total of innate factors which make a nonimmunized host an unsuitable medium in which a given parasite can achieve maximum reproductive and survival rates. It is important to note that this definition includes in a negative sense the lack of suitable foodstuff for a parasite in an immune host or reciprocally the inability to synthesize suitable enzymes by a parasite of low virulence. Also it includes natural antibodies. Whether or not natural antibodies are actually formed as a result of the development and maturation of the host, such as occurs in the human blood isoagglutinins, or whether they represent immune antibodies acquired to other antigenic stimuli, they need not be considered further because they would act in a manner similar to, although generally weaker than, immune antibodies. The definition may be at fault in that it may include parasites so defective as to be unable to survive even in an ideal situation.

As would be expected, innate immunity is to a large extent genetically determined. Thus the inbred strains of mice recently developed for cancer work have also been used to produce different degrees of infections with such trypanosomes as *Trypanosoma cruzi* (see Pizzi et al., 1949). Differences in innate immunity may also occur between the sexes. Bennison and Coatney (1948) found that female chickens possess a lower innate immunity to *Plasmodium gallinaceum* than males.

Many investigators tacitly assume that the ability of a parasite to infect hosts or conversely the innate immunity of a potential host is an all-or-none affair. As analyses of infections progress, however, there seem to be all gradations of susceptibility. Consider for example *Plasmodium gallinaceum,* which causes a highly virulent and frequently lethal malaria in the chicken. Cursory blood examinations would indicate that the goose, duck, guinea fowl, and canary are innately immune to it. Nevertheless a careful study by Huff and Coulston (1946) indicated that these birds comprise a graded series in which innate immunity to the parasite increases. Thus pre-erythrocytic tissue stages develop abundantly in the first three hosts, with the goose suffering a transient parasitemia, the duck a subpatent parasitemia, and the guinea fowl no demonstrable parasitemia, whereas neither pre-erythrocytic nor erythrocytic parasites are found in the canary. Also the adaptability of a parasite often makes it possible to modify host specificity experimentally. A remarkable example is the adaptation to the mouse of the avian malarial parasite *Plasmodium lophurae.* McGhee first found that erythrocytes of the rabbit, pig, mouse, and, to a less extent, man become infected with this plasmodium when inoculated into previously infected chick embryos.

Later he (1951) was able to obtain parasites which could be continuously passed and maintained in the mouse after alternately passing them in chick embryos and mice.

Terzian and his associates (1953 and earlier papers cited by them) have made a detailed study of the innate immunity of the mosquito host *Aëdes aegypti* to *P. gallinaceum*. They found that various sulfonamides, antibiotics, metabolites, and plant and animal hormones modify innate immunity as measured by the number of malaria oöcysts produced. Thus critical concentrations of sulfadiazine, aureomycin, chloramphenicol, penicillin, tyrothricin, streptomycin, para-amino-benzoic acid, ascorbic acid, insulin, thyroxin, indole-3-acetic acid, and β-naphthaoxyacetic acid all decrease innate immunity whereas terramycin, thiamin, niacin, calcium pantothenate, biotin, and ACTH all increase the immunity. Paradoxically some compounds reported effective in decreasing innate immunity of the mosquito act directly on the parasite in the chicken to cure the infection, i.e. to produce the opposite effect. Among these are sulfadiazine (see Wiselogle, 1946) and tyrothricin (L. G. Taliaferro et al., 1944). Terzian et al. (1953) emphasize the complex nature of innate immunity. They suggest that "the innate immunity of a host to a particular parasite is a result of a summation of the interrelations, or of the reciprocal activities, of more than one system rather than the resultant of a single effective factor."

A few instances are known where potential hosts possess highly potent specific innate parasiticidal substances. Man almost certainly owes a part of his innate immunity against the trypanosomes producing disease in the animals of Africa to a nonantibody trypanocidal substance in his serum. This serum component is highly protective and curative against infection with the parasites of animals but does not affect *Trypanosoma gambiense,* the commonest trypanosome producing human sleeping sickness. In addition it does not harm *T. rhodesiense,* which also produces human sleeping sickness and is probably a strain of *T. brucei,* an animal parasite partially adapted to man. This species, however, becomes susceptible to human serum after a few passages in animals. The trypanocidal substance does not pass the placenta although it increases in amount during the later stages of pregnancy. Its formation is associated with the liver and its titer is greatly reduced in certain liver diseases (see review in Culbertson, 1935).

Both innate and acquired immunity may lower the basic rate of reproduction and/or the survival of the progeny. Certain of the animal parasites are unique in that their rate of reproduction and their rate of survival can be measured independently. Those best suited for such studies are the synchronously growing and reproducing malarial parasites (Fig. 1) which can be studied in blood samples without sacrificing the host. Moreover the average number of progeny (merozoites) formed in conjunction with the length of the asexual cycle gives an absolute measure of the reproductive rate independent of the number which die (see L. G. Taliaferro, 1925). An interesting illustration is afforded by the blood infection with *Plasmodium brasilianum* in a Cebus monkey shown in

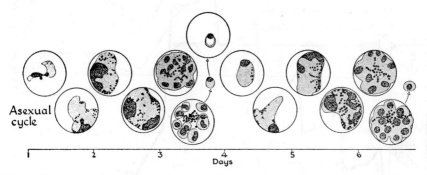

Asexual cycle

Days

Fig. 1. Diagrammatic representation of the synchronous three-day asexual cycle of *Plasmodium brasilianum*. After the parasites are grown, segmentation occurs during a period of about 8 hours on the third day 'at which time each parasite forms about 10 merozoites. After W. H. and L. G. Taliaferro.

Fig. 2. This infection also shows (W. H. and L. G. Taliaferro, 1944) that innate immunity is a highly effective mechanism for reducing the parasite population although the specific mechanisms involved are not known.

The interplay of host immunity and parasite invasiveness can be seen throughout this infection. During the first part of the infection, when innate immunity is operative and acquired immunity has not supervened, the parasites reproduced at a remarkably constant rate, i.e. the asexual cycle was three days long and each segmenter produced a mean of about 10 merozoites during each asexual cycle (Fig. 1 and 2). Provided all the parasites survived, the infection should rise at three-day steps by a factor of 10 according to a geometrical progression. Actually it increased at each segmentation by a factor of about six, while four merozoites died during the passage of the parasite from the red cell of the segmenter to a new red cell. In addition the parasitemia decreased during each intersegmentation period by a factor of three. In other words not only did four parasites die extracorpuscularly but three more died during intracorpuscular development. The net result was a threefold in-

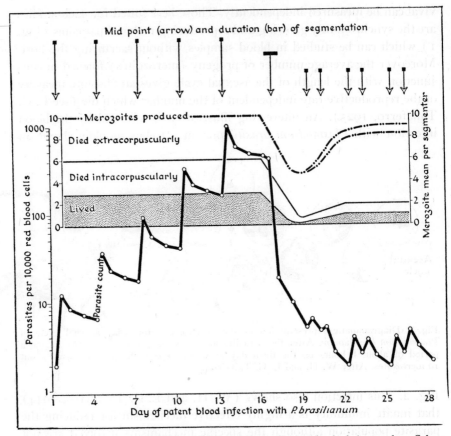

Fig. 2. Population curve of *Plasmodium brasilianum* in the blood of the monkey *Cebus capucinus*. The *merozoites produced,* as shown in Fig. 1, give the number of progeny formed at each segmentation. The number of parasites in the blood indicates what proportion of the progeny died extracorpuscularly and intracorpuscularly and what proportion survived. From data of W. H. and L. G. Taliaferro.

stead of a tenfold increase. All we know about the mechanisms involved in this death of parasites while innate immunity is operating is that the parasites and parasitized erythrocytes are taken up by the macrophages (Plate I), chiefly in the spleen, liver, and bone marrow (Taliaferro and Cannon, 1936). Gingrich (1941), as a result of his study of the effect of blockade on bird malarial infections, believes that such phagocytosis is secondary and involves parasites which succumb because of such possible factors as a plasmodicidal substance in the plasma, the hazard of penetrating new red cells, or the fate of parasites within damaged red cells. On the other hand Becker and his associates (see his review in 1951 and 1952) imply an antibody mechanism for phagocytosis during innate immunity. They have described a sparing substance which inhibits phagocytosis, i.e. lessens innate immunity. This substance they believe to be probably the same as a hemagglutinin-inhibiting principle, related but not identical with the Forssman antigen. The sparing phenomenon is thus pictured as an inhibition phenomenon resulting from the "neutralization" of an opsonic antibody by the Forssmanlike antigen.

A crisis occurs when acquired immunity is superimposed upon innate immunity with a consequent sharp drop in the parasitemia. At this time in the infection shown in Fig. 2, reproduction was markedly delayed, deranged, and reduced in rate (only four merozoites were produced per segmenter) and obviously stunted abnormal parasites circulated in the blood. Death of the parasites at this onset of acquired immunity is unquestionably the result of a specific antibody which produces, first, agglutination (cf. Eaton, 1938) and later, opsonification and phagocytosis (Plate I) of parasitized erythrocytes in the spleen, liver, and bone marrow (Taliaferro and Cannon, 1936, and reviews by Coggeshall, 1943, and Taliaferro, 1949). The antibody, in addition to showing the usual antibody specificity, is frequently associated with phagocytosis of nonparasitized red cells. This phenomenon may result from a generalized damage to the red cell by the infection or from an anti-normal-red-cell element in the antimalarial antibody (cf. Zuckerman, 1945, and Coffin, 1951). The temporary inhibition of reproduction of the parasites which may occur during the crisis is apparently for the most part a secondary effect of the fundamentally parasiticidal antibody which so quickly and effectively removes parasites from the body (see discussion in Taliaferro, 1948). Within a few days, however, the surviving parasites regain their synchronous three-day cycle, normal appearance, and normal production of merozoites. In this particular monkey the post-critical parasites realigned themselves into two broods and the infection maintained itself at an essentially constant level for a long time.

Malaria in general and infections like *P. brasilianum* in particular often exhibit a long-continued chronic course. The question arises, how do parasites survive in a medium containing specific antibodies? A study of the asexual cycle of *P. brasilianum* during the postcritical period indicates that reproduction is continuing at the original rate in spite of a parasitostasis. It follows therefore that of the 8 to 10 parasites which each segmenter produces every three days the equivalent of 7 to 9 are susceptible to the antibody (and innate immunity) and one is not. This difference is phenotypic and not genotypic because otherwise the resistant progeny would immediately repopulate the blood stream with resistant parasites (see discussion of chicken malarias by W. H. and L. G. Taliaferro, 1950, and later discussion of pathogenic trypanosomes). Investigations by Coggeshall (review in 1943) indicate that relapses in another monkey malaria, caused by *P. knowlesi,* with temporary increases of parasitemia are associated with a diminution in circulating antibody. Furthermore, Coggeshall (1943) found that a specific immune serum would protect a normal monkey against an infection with the parasites surviving in the same serum. Therefore the progeny of the surviving parasites possess the original antigenic structure of the parasites which stimulated the formation of the original antibody.

Among the blood-inhabiting trypanosomes, population statistics can be followed as accurately as among the malarial parasites. The rate of reproduction, however, can only be measured relatively during the course of the infection in terms of changes (1) in variability in size due to fission and growth or (2) in the percentage of division stages. With these methods it has been possible to correlate several types of infection with the antibody response.

Certain of the pathogenic trypanosomes (e.g. *Trypanosoma equinum,*

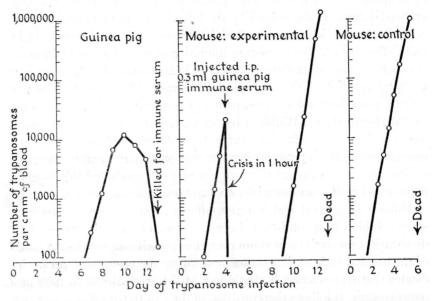

Fig. 3. The *in vivo* action of a trypanolysin (immune serum) obtained from a guinea pig infected with *Trypanosoma equinum* (left graph). Two mice were infected with the same strain of parasites used to infect the guinea pig. One (middle graph), after treatment with 0.3 ml immune serum per 20 gm. mouse, had a crisis but later had a fatal relapse. The untreated control (right graph) showed a progressive fatal infection. Redrawn from Taliaferro and Johnson.

[164]

T. equiperdum, T. brucei, T. gambiense, etc.) reproduce at a constant rate in a series of experimental hosts (W. H. and L. G. Taliaferro, 1922). In mice and at times in rats the infection rises essentially logarithmically until death occurs (Fig. 3, Mouse, control). There seems to be no evidence of antibody formation and innate immunity can not be estimated with the measures at hand. In other hosts such as the guinea pig (Fig. 3) the parasitemia successively decreases due to the production of typical trypanolysins. Trypanolytic crises rarely eradicate the infection, however, because a few resistant trypanosomes continue to reproduce and repopulate the blood stream until a lysin for the relapse strain is formed (Fig. 3, Mouse, experimental). Ability of trypanosomes to acquire new heritable antigens prolongs the infection in spite of the successive production of new lysins, but eventually most hosts die (see reviews by Taliaferro, 1926, of work by Ehrlich and his co-workers, Rodet and Vallet, Massaglia, Levaditi and Mutermilch, and Ritz; Harrison, 1947; and Inoki et al., 1952).

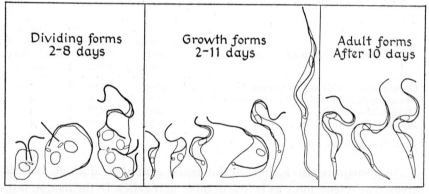

Fig. 4. *Trypanosoma lewisi* in the rat. Dividing forms occur early in the infection, growth forms from the second to the eleventh day, and nonreproducing forms later. No reproduction of the trypanosomes occurs in an experimental rat when given nonreproducing adults and ablastic serum, i.e. serum from a rat in whose blood trypanosomes have stopped reproducing. Adapted from Taliaferro and Coventry.

A unique antibody mechanism is the specific inhibition of growth and reproduction, which so far has been unequivocally demonstrated only in the *Trypanosoma lewisi* group of nonpathogenic trypanosomes. As long ago as 1899 Rabinowitsch and Kempner found that *T. lewisi,* the most studied species, undergoes reproduction by division only during the first part of its infection in the rat and thereafter remains in the blood as a nonreproducing adult form (Fig. 4). In 1924 the author showed that this supposed inherent life cycle is due to the formation of an antibody, ablastin, which inhibits reproduction of the parasites without any apparent effect on their general vitality, mobility, or infectivity. Moulder (1948) has found that trypanosomes inhibited by ablastin have a higher oxygen uptake and a higher respiratory quotient but a lower glucose utilization (Table I). He has adduced persuasive evidence that ablastin acts by inhibiting the oxidative assimilation of glucose in a manner similar to the action of azide or dinitrophenol in concentrations too low to reduce the rate of oxygen consumption. Under these conditions the parasites would be prevented from oxidatively assimilating glucose for the synthesis of new protoplasm but would be able to oxidize glucose for maintenance as nonreproducing organisms. Nonreproducing trypanosomes may persist in the rat for varying lengths of time up to more than a year but are eventually destroyed by specific lysins. Any trypanosomes resistant to lysins, however, do not repopulate the blood because they can not reproduce in the presence of ablastin (Coventry, 1930). Ablastin-resistant organisms have not as yet been found.

TABLE I. Oxidative metabolism of *Trypanosoma lewesi*.

Reproductive state of trypanosomes	Dividing and growing	Non-dividing adults	Day of infection on which change occurred
Oxygen uptake μM/10^9 trypanosomes/2 hr.	45.00	62.00	5
Glucose utilization μM/10^9 trypanosomes/2 hr.	42.00	25.00	4
Oxygen/glucose ratio	1.03	2.78	4
Respiratory quotient	0.74	0.91	4

Data from Moulder (1948).

An ablastin is also formed in the mouse against *T. duttoni*. It is decreased by splenectomy as is true of other antibodies. In the mouse, however, an *innate* ablastic principle is also associated with the spleen but so far has neither been identified as an innate antibody nor been passively transferred as a serum property (Taliaferro, 1938).

Another type of antibody action is seen during infection of the rat with the small hookwormlike parasite *Nippostrongylus muris*. Rats acquire this infection from infective larvae which penetrate the skin, migrate via the blood to the alveoli after about 20 hours, and make a second migration between the second and third day via the trachea, buccal cavity, and esophagus to the intestine, where they colonize the anterior third of the jejunum. During the second week the females lay large numbers of eggs which pass out with the feces and develop eventually into infective larvae. Then during the third week most of the adults are swept out of the intestine as a result of acquired immunity and few if any eggs are passed in the feces thereafter.

Acquired immunity increases as successive reinfections take place. As shown by Taliaferro and Sarles (1939) it arises as a consequence of the presence of antibodies formed in response to the secretions and excretions which pass from the mouth, anus, and excretory pore of the worm. These antibodies act as precipitins to cause precipitates to form

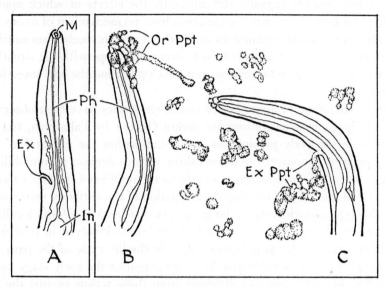

Fig. 5. Infective larvae of *Nippostrongylus muris* after 20 hours (A) in normal serum and (B and C) in immune serum. Note no precipitates in normal serum and copious precipitates at the mouth and excretory pore and in the medium in immune serum. Ex, excretory pore; Ex Ppt, excretory pore precipitate; In, intestine; M, mouth; Or. Ppt, oral precipitate. Adapted from Sarles.

at all orifices and throughout the gut of feeding larvae both *in vitro* (Fig. 5) and *in vivo* in the skin and lungs (Plate II). Antibodies—probably the same precipitins—also cause larvae to be stunted and more or less immobilized. Furthermore the intense antigen-antibody reactions in the tissues produce tissue damage and set up inflammation which in turn hastens the formation of nodules (Plate II) around the immobilized parasites. Some worms die in these nodules and the disintegrating worms are removed by macrophages. Others escape from the nodules and resume their migration but are stunted and the females lay fewer eggs than normally. In the intestine the worm feeds by piercing the mucosa and extracting materials from the lamina propria. Just as in the skin and lungs, this process does little damage until acquired immunity super-

venes. In the immune rat, however, material extruded from the mouth
sets up inflammatory processes within the tissue. In addition ingestion
of antibody probably immobilizes the worm to the extent that it can
not withstand peristalsis and is forced out of the gut.

It should be noted that immunity to *Nippostrongylus* often involves
a definite inhibition of reproduction because the stunted females lay so
few eggs. This as in malaria is not a specific inhibition but is the result
of a fundamentally parasiticidal antibody, the effects of which appear
slowly. As far as is known, the parasites have no mechanism of overcom-
ing the effects of this parasiticidal antibody. A genetic mechanism causing
antibody-resistant strains would not successfully cope with the situation
because a second generation does not develop beyond the egg stage in a
given host.

Infective larvae of a number of nematode species have been observed
in vitro in the presence of immune serum (review by Taliaferro, 1943).
Just as in the body precipitates are found within the gut and at the
body orifices. Moreover the worms show various degrees of immobilization and death. Of special interest is the fact that Oliver-González (1943)
found that the antigen inciting the formation of the antibody responsible
for this type of action on the larvae of *Ascaris lumbricoides* is localized
in the egg of the adult worm. In fact antibodies involved in immunity
to worms may differ at different stages in the life cycle of the parasite.
Campbell (1938) found antibodies acting against the early stage (presumably the onchospheres) different from those acting against the developed cysticerci during the larval development of *Taenia taeniaeformis* in rats. The complete absence of cysticerci indicates that an early
stage is involved in the passive immunity shown in Fig. 6. Similarly,
Oliver-González (1941) reported different antibodies acting in trichinosis on the adult and newly deposited larvae and on the developed muscle trichinae.

Although we have excluded ectoparasites from consideration, it should
be mentioned that acquired immunities to the larvae of the myiasis-
producing flies (Blacklock et al., 1930) and to ticks (Trager, 1939) have
been demonstrated. The antibody mechanisms act fundamentally as during the intestinal stage of nematode infections.

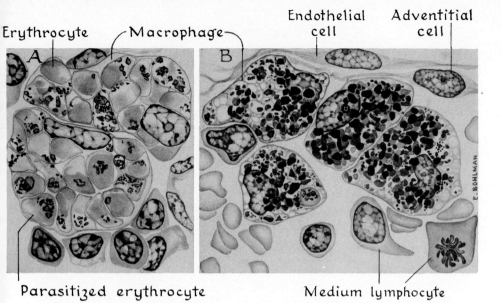

Erythrocyte — Macrophage — — Endothelial cell — Adventitial cell

Parasitized erythrocyte   Medium lymphocyte

PLATE I. Splenic macrophages containing phagocytosed material from a Cebus monkey infected with *P. brasilianum*. A, intense antibody-induced phagocytosis during the crisis of the infection; B, malarial pigment remaining after digestion of the parasites and red cells, which itself is digested within several months. After Taliaferro and Cannon.

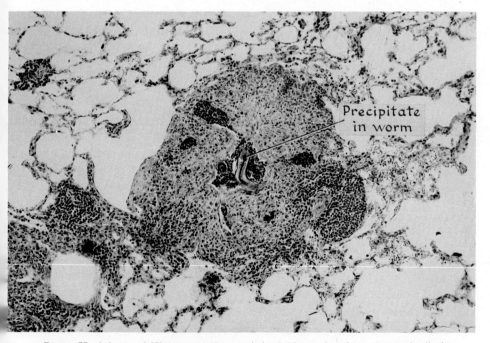

Precipitate in worm

PLATE II. A larva of *Nippostrongylus muris* in the lung of an immune rat. Antibody immobilized the larva and reacted with parasite antigens to form a cap of precipitate over the mouth and a core of precipitate throughout the intestine. Inflammatory processes resulted in the formation of a granulomatous nodule around the larva. Adapted from Taliaferro and Sarles.

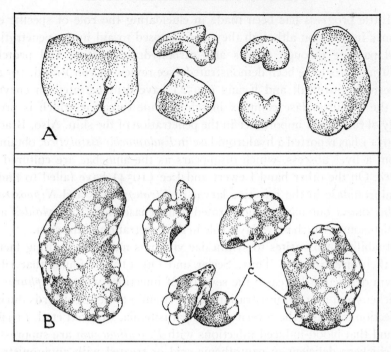

Fig. 6. The protective action of immune serum against infection with the cysticercus of the cat tapeworm *Taenia taeniaeformis* as seen in the lobes of the liver from two rats. Rat A received 1,000 onchospheres plus 1 ml immune serum per 25 gm. body weight, i.e. serum from a previously infected rat, whereas rat B received the same number of onchospheres plus normal serum. No cysticerci occurred in rat A but 256 living and 13 dead ones were found in rat B. This antibody probably kills onchospheres; another kills developed cysticerci. Redrawn from Miller and Gardiner.

II. FACTORS PRIMARILY ASSOCIATED WITH PARASITE INVASIVENESS AND VIRULENCE

Under invasiveness of the parasite are included all of the factors which permit the parasite to invade and colonize the host. Parenthetically it should be pointed out that virulence, or the power to produce disease, is broadly considered to be a combination of invasiveness and toxigenicity of the parasite. Present experimental evidence indicates that the toxigenicity of animal parasites is so low that it can be ignored and that virulence can be considered solely as invasiveness. Invasive factors are or rest upon specific biochemical entities for the most part. Nevertheless they comprise such a heterogeneous collection of items that specific substances can rarely be associated with specific activities of the parasite.

[169]

Little progress has been made in elucidating the role of specific enzymes in invasion although they are supposed to aid in the penetration and passage through tissues and carbohydrate-splitting and protein-splitting ones have been demonstrated (see review in von Brand, 1952). Recently, Stirewalt and Evans (1952) have demonstrated an enzyme of the hyaluronidase complex in schistosome cercariae which is probably of functional importance in the penetration of the skin. Also, Bradin (1951) has reported a hyaluronidase in *Endamoeba histolytica,* obtained from a liver abscess, which disappears as the amoebae are cultured *in vitro.* On the other hand Lewert and Lee (1954) have failed to find a hyaluronidase in the infective larvae of *Strongyloides* and *Nippostrongylus muris* but have demonstrated collagenases in *Strongyloides* and schistosomes which may play a role in the penetration of the skin.

Inability of parasites to synthesize vitamins may be a limiting factor in their invasiveness. Thus, Seeler and Ott (1944) found that riboflavin deficiency decreases the severity of infections with *P. lophurae* in chickens and that administration of the vitamin to deficient birds during infection increases the severity of the parasitemia. Brackett et al. (1946) found that blood-induced infections with *P. gallinaceum* are suppressed in chickens deficient in pantothenic acid or treated with antagonists of pantothenic acid (see also Wiselogle, 1946). They believe that their findings are an expression of the pantothenic acid requirement of the erythrocytic stages of the parasite. It is noteworthy that in pantothenic acid-deficient chickens the rate of reproduction of the parasite, i.e. the mean number of merozoites formed per segmenter, was reduced to about the same degree as found by W. H. and L. G. Taliaferro (1950) during the onset of acquired immunity. The foregoing results are in marked contrast to those obtained with biotin deficiency, which is associated with a heightened parasitemia and which may act indirectly to lower innate immunity of the host (Trager, 1947).

Invasiveness is closely related to the elective localization of parasites to certain cells, tissues, or organs. This condition is almost certainly associated in part with the availability of suitable nutrients or specific enzyme systems which the parasite can utilize. In spite of much speculation little is known about the fundamental mechanisms involved. Consider for example the localization of the erythrocytic stages of the malarial parasites within red cells. There is excellent reason to assume that these parasites possess an enzyme system which splits hemoglobin into heme and globin and that they then utilize globin as a nitrogen source and discard heme (possibly not all) which accumulates as malarial pigment

[170]

(Fig. 1 and Plate I) (see reviews by Moulder, 1948; Groman, 1951; and von Brand, 1952). The exoerythrocytic stages on the other hand must obviously obtain their nitrogenous compounds from other sources since they do not have access to hemoglobin. These facts leave open to question whether hemoglobin-splitting enzymes are primary or secondary to the localization of the parasite within the red cell.

It is worthy of note that localization of a given parasite can be modified. Thus adults of the nematode *Nippostrongylus muris* localize in the anterior third of the jejunum of the rat. As local immunity is acquired at their preferred site, they appear to localize farther and farther posteriorly (cf. Taliaferro and Sarles, 1939).

III. MUTUALISM

So far I have emphasized some of the factors which, on the one hand, permit parasite invasion and, on the other hand, hold invasion down sufficiently to prolong the host-parasite association. It has been tacitly assumed that parasite invasion is harmful or at best a matter of indifference to the host. In true mutualism (sometimes termed symbiosis) where both host and parasite are benefited, however, an optimal invasion may have survival importance to the host. A remarkable example of this is seen in the work by Cleveland, Hungate, and others on the intestinal flagellates of certain wood-eating termites. In summarizing this work von Brand (1952) states, "The flagellates are benefited by finding shelter in a habitat with the required low oxygen tension and they are provided by termites with small ingestible wood particles as well as with nitrogenous material. In return, the flagellates furnish the insects degradation products of cellulose which the latter can absorb and utilize in their energy metabolism."

In these mutualistic associations the parasite may respond to physiological regulatory mechanisms of the host. An illuminating example is the discovery by Cleveland that molting of the wood-feeding roach *Cryptocercus punctulatus* is associated with sexual cycles in several of the flagellates of its hind gut. The sexual behavior of the flagellates varies but is apparently induced by the molting hormone which is produced by the prothoracic glands of the roach. Cleveland (see 1949) doubts, however, that the hormone is responsible for fertilization and meiosis. Since the induction of sexual phenomena might well aid in the survival of the infection by increasing variation and producing variants better suited to survive, the phenomenon could be a factor in invasiveness as in the case of antibody resistance of trypanosomes.

[171]

Even in the ideal host-parasite relationship seen in the termite-flagellate association, various factors must hold unlimited reproduction of the parasites in check. Andrew and Light (1929) observed that there are three phases in the intestinal flagellate infection of *Termopsis angusti-collis:* (1) a period of rapid flagellate multiplication following reinfection of the recently molted termite, (2) a long period between molts when reproduction and death of the protozoa are low or practically nil, and (3) a short period of protozoan reduction by elimination or death during molting. Hungate (1941) suggests that the lack of reproduction between molts and the rapid reproduction just after molting are due to the availability of nitrogenous materials necessary for protozoan growth. Restraint of the flagellates during the intermolting periods from a lack of nitrogenous food would fall within Ehrlich's concept of athreptic immunity. It should be stressed, however, that limiting foodstuff may be specific compounds. Such a specificity is indicated by Trager's (1934) finding that *Trichomonas termopsidis,* one of the flagellates under consideration, can only utilize cellulose as a carbon source in culture.

IV. SUMMARY

In conclusion I should like to reemphasize the fact that successful endoparasitism rests on a complex shifting dynamic balance between specific physiological processes of the host and those of the parasite. Of the specific factors involved in this balance, most is known about antibodies. Among other factors, potential hosts obviously possess many innate physiological mechanisms which, in conjunction with the inadequacies of the parasites, completely or partially restrict parasitism. Among the likely inadequacies of the parasite are its inability to synthesize vitamins and its lack of specific enzymes. Conversely the potential host may be protected by an insufficient concentration of vitamins or the lack of suitable enzyme substrates.

In the case of partially restricted parasitism which is necessary in the long survival of both parasite and host, the host frequently responds by forming antibodies. In infections with *Trypanosoma lewisi,* the antibody, ablastin, inhibits reproduction of the parasite without restricting motility, infectivity, or maintenance metabolism of the parasite. In the mutualism between termites and their intestinal flagellates, reproduction of the latter may be controlled by the availability of nitrogenous compounds. Most antibodies do not restrict parasites by an inhibition of reproduction over long periods but act to kill and remove them quickly from the body (Taliaferro, 1948). Nevertheless parasitism under such

[172]

conditions can be prolonged in several ways. For example precipitins, which inhibit growth and reproduction of the nematode worms, act so slowly that parasitism is greatly prolonged. By another mechanism some trypanosomes prolong their parasitism in the presence of highly effective lysins by virtue of the variation and inheritance of specific antigens. The malarial parasites, although susceptible to opsonins and parasiticidal antibodies, show a marked phenotypic variability. Thus over long periods a few parasites of each brood survive and keep the infection going.

BIBLIOGRAPHY

Andrew, B. J., and S. F. Light. 1929. Natural and artificial production of so-called "mitotic flares" in the intestinal flagellates of *Termopsis angusticollis*. *Univ. Calif. Pub. Zool. 31,* 433-440.

Baer, J. G. 1951. *Ecology of Animal Parasites*. University of Illinois Press, Urbana.

Becker, E. R., T. M. Schwink, and R. M. Prather, Jr. 1951. Hemagglutinin-inhibiting property of duck plasma exhibited in agglutination reactions involving duck erythrocytes and plasma of chicks recovered from lophurae malaria. *J. Inf. Dis. 89,* 95-102.

Becker, E. R., T. M. Schwink, and R. T. Probst. 1952. The nature of the sparing phenomenon. 1. Activity and nature of hemagglutinin-inhibitor. *Iowa State Col. J. Sci. 27,* 79-90.

Bennison, B. E., and G. R. Coatney. 1948. The sex of the host as a factor in *Plasmodium gallinaceum* infections in young chicks. *Science 107,* 147-148.

Blacklock, D. B., R. M. Gordon, and J. Fine. 1930. Metazoan immunity : a report of recent investigations. *Ann. Trop. Med. Hyg. 24,* 5-54.

Brackett, S., E. Waletzky, and M. Baker. 1946. The relation between pantothenic acid and *Plasmodium gallinaceum* infections in the chicken and the antimalarial activity of analogues of pantothenic acid. *J. Parasit. 32,* 453-462.

Bradin, J. L. 1951. Hyaluronidase production by *Endamoeba histolytica. J. Parasit. Suppl. 37,* 10-11.

von Brand. T. 1952. *Chemical Physiology of Endoparasitic Animals*. Academic Press, New York.

Campbell, D. H. 1938. The specific protective property of serum from rats infected with *Cysticercus crassicollis. J. Immun. 35,* 195-204.

Cleveland, L. R. 1947. Sex produced in the protozoa of *Cryptocercus* by molting. *Science 105,* 16-18.

Cleveland, L. R. 1949. Hormone-induced sexual cycles of flagellates *J. Morph. 85,* 197-296.

Cleveland, L. R., S. R. Hall, E. P. Sanders, and J. Collier. 1934. The wood-feeding roach *Cryptocercus,* its protozoa, and symbiosis between protozoa and roach. *Mem. Am. Acad. Arts Sci. 17,* 185-342.

Coffin, G. S. 1951. Passive immunization of birds against malaria. *J. Inf. Dis.* *89*, 8-15.

Coggeshall, L. T. 1943. Immunity in malaria. *Medicine 22*, 87-102.

Coventry, F. A. 1930. The trypanocidal action of specific antiserums on *Trypanosoma lewisi in vivo. Am. J. Hyg. 12*, 366-380.

Culbertson, J. T. 1935. Trypanocidal action of normal human serum. *Arch. Path. 20*, 767-790.

Eaton, M. D. 1938. The agglutination of *Plasmodium knowlesi* by immune serum. *J. Exp. Med. 67*, 857-870.

Gingrich, W. D. 1941. The role of phagocytosis in natural and acquired immunity in avian malaria. *J. Inf. Dis. 68*, 37-45.

Groman, N. B. 1951. Dynamic aspects of the nitrogen metabolism of *Plasmodium gallinaceum in vivo* and *in vitro. J. Inf. Dis. 88*, 126-150.

Harrison, J. A. 1947. Antigenic variation in protozoa and bacteria. *Ann. Rev. Microbiol. 1*, 19-42.

Huff, C. G., and F. Coulston. 1946. The relation of natural and acquired immunity of various avian hosts to the cryptozoites and metacryptozoites of *Plasmodium gallinaceum* and *Plasmodium relictum. J. Inf. Dis. 78*, 99-117.

Huff, C. G. 1951. Observations on the pre-erythrocytic stages of *Plasmodium relictum, Plasmodium cathemerium,* and *Plasmodium gallinaceum* in various birds. *J. Inf. Dis. 88*, 17-26.

Hungate, R. E. 1941. Experiments on the nitrogen economy of termites. *Ann. Entomol. Soc. Am. 34*, 467-489.

Inoki, S., T. Kitaura, Y. Kurogochi, H. Osaki, and T. Nakabayasi. 1952. Genetical studies on the antigenic variation in *Trypanosoma gambiense. Jap. J. Gen. 27*, 85-92.

Lewert, R. M., and C. Lee. 1954. Studies on the passage of helminth larvae through host tissues. *J. Inf. Dis. 95*, 13-51.

McGhee, R. B. 1951. The adaptation of the avian malaria parasite, *Plasmodium lophurae,* to a continuous existence in infant mice. *J. Inf. Dis. 88*, 86-97.

Moulder, J. W. 1948. Changes in the glucose metabolism of *Trypanosoma lewisi. J. Inf. Dis. 83*, 42-49.

Moulder, J. W. 1948. The metabolism of malarial parasites. *Ann. Rev. Microbiol. 2*, 101-120.

Oliver-González, J. 1941. The dual antibody basis of acquired immunity in trichinosis. *J. Inf. Dis. 69*, 254-270.

Oliver-González, J. 1943. The antigenic analysis of the isolated tissues and body fluids of the round-worm, *Ascaris lumbricoides var. suum. J. Inf. Dis. 72*, 202-212.

Pizzi, T., M. Agosín, R. Christen, G. Hoecker, and A. Neghme. 1949. Estudios sobre immunobiología de las enfermedades parasitarias. 1. Influencia de la constitución genética en la resistencia de las lauchas a la infección experimental por *Trypanosoma cruzi. Bol. Inform. Parasit. Chilenas 4*, 48-49.

Rabinowitsch, L., and W. Kempner. 1899. Beitrag zur Kenntniss der Blutparasiten, speciell der Rattentrypanosomen. *Zeits. Hyg. Infekt. 30,* 251-294.

Sarles, M. P. 1938. The *in vitro* action of immune rat serum on the nematode, *Nippostrongylus muris. J. Inf. Dis. 62,* 337-348.

Seeler, A. O., and W. H. Ott. 1944. Effect of riboflavin deficiency on the course of *Plasmodium lophurae* infection in chicks. *J. Inf. Dis. 75,* 175-178.

Stirewalt, M. A., and A. S. Evans. 1952. Demonstration of an enzymatic factor in cercariae of *Schistosoma mansoni* by the streptococcal decapsulation test. *J. Inf. Dis. 91,* 191-197.

Taliaferro, L. G. 1925. Infection and resistance in bird malaria with special reference to periodicity and rate of reproduction of the parasite. *Am. J. Hyg. 5,* 742-789.

Taliaferro, L. G., F. Coulston, and M. Silverman. 1944. The antimalarial activity of tyrothricin against *Plasmodium gallinaceum. J. Inf. Dis. 75,* 179-211.

Taliaferro, W. H. 1924. A reaction product in infections with *Trypanosoma lewisi* which inhibits the reproduction of the trypanosomes. *J. Exp. Med. 39,* 171-190.

Taliaferro, W. H. 1926. Host resistance and types of infections in trypanosomiasis and malaria. *Quart. Rev. Biol. 1,* 246-269.

Taliaferro, W. H. 1938. Ablastic and trypanocidal antibodies against *Trypanosoma duttoni. J. Immun. 35,* 303-329.

Taliaferro, W. H. 1943. Antigen-antibody reactions in immunity to metazoan parasites. *Proc. Inst. Med. Chicago 14,* 358-368.

Taliaferro, W. H. 1948. The inhibition of reproduction of parasites by immune factors. *Bact. Rev. 12,* 1-17.

Taliaferro, W. H. 1949. Immunity to malaria infections, in M. F. Boyd, *Malariology 2,* 935-965. W. B. Saunders Co., Philadelphia.

Taliaferro, W. H., and P. R. Cannon. 1936. The cellular reactions during primary infections and superinfections of *Plasmodium brasilianum* in Panamanian monkeys. *J. Inf. Dis. 59,* 72-125.

Taliaferro, W. H., and M. P. Sarles. 1939. The cellular reactions in the skin, lungs and intestine of normal and immune rats after infection with *Nippostrongylus muris. J. Inf. Dis. 64,* 157-192.

Taliaferro, W. H., and L. G. Taliaferro. 1922. The resistance of different hosts to experimental trypanosome infections, with especial reference to a new method of measuring this resistance. *Am. J. Hyg. 2,* 264-319.

Taliaferro, W. H., and L. G. Taliaferro. 1944. The effect of immunity on the asexual reproduction of *Plasmodium brasilianum. J. Inf. Dis. 75,* 1-32.

Taliaferro, W. H., and L. G. Taliaferro. 1950. Reproduction-inhibiting and parasiticidal effects on *Plasmodium gallinaceum* and *Plasmodium lophurae* during initial infection and homologous superinfection in chickens. *J. Inf. Dis. 86,* 275-294.

Terzian, L. A., N. Stahler, and H. Miller. 1953. A study of the relation of antibiotics, vitamins and hormones to immunity to infection. *J. Immun. 70,* 115-123.

Trager, W. 1934. The cultivation of a cellulose-digesting flagellate, *Trichomonas termopsidis,* and of certain other termite protozoa. *Biol. Bull. 66,* 182-190.

Trager, W. 1939. Acquired immunity in ticks. *J. Parasit. 25,* 57-81.

Trager, W. 1947. The relation to the course of avian malaria of biotin and a fat-soluble material having the biological activities of biotin. *J. Exp. Med. 85,* 663-683.

Wiselogle, F. Y. 1946. *A Survey of Antimalarial Drugs.* Vol. 2. J. W. Edwards, Ann Arbor.

Zuckerman, A. 1945. *In vitro* opsonic tests with *Plasmodium gallinaceum* and *Plasmodium lophurae. J. Inf. Dis. 77,* 28-59.

IX. COMPATIBILITY AND NON-COMPATIBILITY IN TISSUE TRANSPLANTATION

BY HARRY S. N. GREENE[1]

THE transplantation of tissues depends on a variety of factors, the most important being the constitutional relationship of the donor and recipient, the status of the transplant, and the site of transplantation. The constitutional factors concerned are generally considered to be genetic in nature and of such significance that differences in strain or species constitute barriers to the exchange of tissues. There is evidence, however, that other nongenetic characteristics may be of equal importance, and further that the success or failure of transfers performed under identical genetic circumstances may be conditioned by the status of the transplant or the site of transplantation. The interrelationship of the factors determining transplantability has been of primary interest in this laboratory, and the present discussion is based on data derived from the study of those factors.

In essence the investigation consisted of a comparison of the transplantation reactions of various tissue states in hosts of different constitutional status. The tissues investigated included normal adult tissue, normal embryonic or fetal tissue, hyperplastic or benign tumor tissue, precancerous tissue, and cancer. These tissues were transferred to other regions in the same host (autologous), to unrelated normal and abnormal individuals of the same species (homologous), and to animals of alien species (heterologous). The anterior chamber of the eye, the brain, the subcutaneous space, the testicle, and various internal organs were used as transplantation sites.

I. ADULT TISSUES

Adult tissues generally survive transfer back elsewhere in the primary host or to unrelated animals of the same species, but heterologous transfer is invariably unsuccessful.

[1] Department of Pathology, Yale University School of Medicine. The original investigations reported in this chapter were supported in part by grants from the Medical Research and Development Board, Office of the Surgeon General, Department of the Army, under Contract No. DA-40-007md-30; the Jane Coffin Childs Memorial Fund for Medical Research; the National Cancer Institute; National Institutes of Health, Public Health Service; and the American Cancer Society upon recommendation by the Committee on Growth of the National Research Council.

[177]

The anterior chamber of the eye and the brain are the more suitable sites for homologous transplantation, and takes of such organs as the thyroid, ovary, lung, kidney, salivary gland, and skin occur with high frequency. Transfer of the adrenal, pituitary, and testicle results in a lesser number of takes, while the lowest incidence is associated with the transplantation of brain tissue. Despite numerous attempts, adult liver tissue has not been successfully transplanted. The subcutaneous space is less satisfactory as a transplantation site; but despite a much lower incidence of takes, growth has been obtained on transfer of all the tissues noted above with the exception of the liver, brain, and pituitary.

Transplants of adult tissue are well vascularized but show little increase and persist indefinitely. Brain transplants consist of a proliferation of glial elements and ganglion cells are not present, but other tissues are histologically intact (Plate I, Fig. 1-6; Plate II, Fig. 7).

The integrity of the transplants is demonstrated by their response to functional stimuli and to infection. The ability to carry out normal functions is most readily shown in the case of endocrine transplants. The presence of adult pituitary tissue in the eyes of hypophysectomized rats prevents the occurrence of signs of pituitary deficiency, and such animals remain normal in appearance and behavior. The atrophy of gonads characteristic of hypophysectomized rats does not occur, and fertile matings have been obtained. Thyroidectomized rabbits bearing anterior chamber transplants of adult thyroid gland have been held for observation for several years, and throughout this period the serum iodine values have been maintained at normal levels. Transplants of adult rabbit ovary respond to the injection of gonadotropic hormone, and the reaction occurs with such rapidity that the increase in size can be used in practice to determine the presence of the hormone in the urine of pregnant women. The function of nonendocrine organs is difficult to demonstrate, but an indication of activity in anterior chamber transplants of kidney is given by the presence of phenosulphophthaline in the aqueous humor after intramuscular injection. The response to infectious agents is also a criterion of the vitality and integrity of the transplant, and the maintenance of the ability to react specifically is readily shown in the case of the skin. The skin of the rabbit is susceptible to infection with the Shope virus and responds with the production of a papillomatous growth in from 17 to 30 days. This ability is retained by the transplants, and fragments of adult skin infiltrated with the virus and transplanted to the brains of other rabbits develop the characteristic lesion (Plate II, Fig. 8).

A point of interest in relation to skin transplantation concerns the fact that the immunity reactions characteristic of the intact animal are not shared by isolated fragments of the epidermis. Rabbits bearing Shope papillomas develop antibodies in sufficient titer to render the animals immune to reinfection. If fragments of skin are removed from such animals, washed in saline, inoculated with the virus, and transplanted to the brains of normal rabbits, typical papillomas arise in the transplant. Further, the papillomas occur when immune animals are used as recipients. In fact skin of an immune animal so treated develops papillomas when transplanted autologously (Plate II, Fig. 9). The point to be emphasized is that the adverse inflammatory reactions so frequently encountered in the homologous transplantation of adult tissues may arise from contaminating substances derived from body fluids rather than from "incompatibility factors" resident in the cells themselves.

II. EMBRYONIC TISSUES

Embryonic tissues, unlike adult tissues, are transplantable heterologously as well as autologously and homologously (Greene, 1943; 1945). The ability to grow in foreign species disappears during the fifth month of gestation in man and at analogous periods of embryonic development in other species.

A. Homologous transplantation. All of the tissues of the embryo with the exception of the liver and the placenta have been successfully transplanted as isolated units in adult animals. The eye and the brain have many advantages as transplantation sites, but all of the tissues except the brain will survive and grow in other regions.

Isolated organs increase rapidly in size after transfer, but progressive growth ceases with differentiation and organization, and the transplants persist indefinitely as miniature replicas of their adult counterparts. However, if two embryonic organs are transplanted simultaneously to the same host, growth continues beyond differentiation and organization, and an approximation of adult size may be attained. The influence of a second embryonic part is also evident in the organization of some organ transplants, notably the adrenal. If an isolated embryonic adrenal is transferred to the anterior chamber of the eye, growth occurs, but the resulting organ shows no architectural pattern and consists only of an unorganized mass of adrenal cells. In contrast when a second embryonic part is added, the adrenal cells form an organoid structure with glomerulosa, fasciculata, and reticularis zones simulating in detail the adult gland (Plate II, Fig. 10). A further influence of a second embryonic part is manifested in liver and placental transplants. Despite the fact that neither of these tissues can be transplanted as isolated organs, growth occurs if either is transferred in association with some other embryonic tissue (Plate II, Fig. 11-12). The nature of mechanism of the influence of one embryonic part on another in the regulation of growth and organization is unknown, but its existence is of particular interest as a manifestation of biological dependence—a phenomenon to which further reference will be made in paragraphs concerned with the transplantation of cancer.

The mature transplants resemble adult organs in histological detail (Plate III, Fig. 13-18). Further, the development of function and specific susceptibilities are unchanged in the new location. Anterior chamber transplants of pancreas produce insulin in diabetic mice, and transplanted adrenals maintain adrenalectomized guinea pigs in health. The transplantation of embryonic human adrenals to patients with Addison's disease has also been successfully accomplished, and in such cases the symptoms of the disease disappear and the patients require no therapy (Plate IV, Fig. 19).

The susceptibility to infection characteristic of the adult tissue is also manifested by the embryonic transplants, and in a number of cases embryonic tissues have been found susceptible to infectious agents to which their adult counterparts are completely resistant. Embryonic rabbit epidermis is highly susceptible to the Shope papilloma virus, and transplants have been of value in investigation of the disease (Greene, 1953). Similarly the susceptibility to other infectious agents natural to the species is generally greater than that of adult tissues, but a characteristic of more biological and clinical significance is the increased susceptibility range.

In some instances the heterologous agent survives and multiplies in the embryonic transplant without producing the specific lesion, while in others the characteristic anatomical changes of the disease are reproduced. The Rous chicken sarcoma virus, the fowl pox virus, and the human influenza virus are examples of the former category. Embryonic mouse and rabbit lung are both susceptible to infection with the human influenza virus; the virus multiplies in the transplants and can be maintained in the alien species without need of the long adaptive procedures necessary for growth in the adult mouse brain. Both the Rous chicken sarcoma virus and the fowl pox virus survive and multiply in homologous transplants of a variety of embryonic tissues. The transplants are not changed morphologically by the virus, but the injection of filtrates of the transplants into chickens results in the formation of typical lesions. Hansen's bacillus, the infectious agent of human leprosy, is inconstant in its action on the transplant. Although all attempts to cultivate this bacillus in experimental animals or in artificial media have been unsuccessful, it survives and multiplies in embryonic guinea pig tissues transplanted to the adult guinea pig brain. In the majority of cases the transplants are unaffected, but occasionally the specific granulomatous lesions of leprosy are found. In contrast the Shope papilloma virus induces characteristic changes. Adult rat epidermis, like the skin of all species except the rabbit, is completely resistant to this virus, yet embryonic rat skin is susceptible and infected transplants growing in adult rat brain respond with the formation of the typical lesion (Plate IV, Fig. 20) (Greene, 1953).

The susceptibility range of embryonic tissue and the ability to react to some infections in a specific manner suggested the application of the procedure to a study of the possible virus etiology of several human diseases. Preliminary investigations in the case of one disease have been promising. Boeck's sarcoid is a disorder of unknown etiology in man and is characterized by a granulomatous lesion resembling tuberculosis. Embryonic mouse thymic tissue was bathed in a filtrate of an involved lymph node before transplantation to the brains of adult mice, and when the animals were killed 3 months later the growing transplant contained the specific granulomatous changes of Boeck's sarcoid (Plate IV, Fig. 21-22). It should be emphasized that this study involves a single case and the results require confirmation. The experiment is cited only as an illustration of the possible application of the finding to the investigation of some human diseases.

A further feature of embryonic transplants of biological interest concerns their high susceptibility to carcinogenic agents (Greene, 1945). The application of methylcholanthrene to the skin of adult mice is followed by the appearance of epidermoid carcinomas after the passage of 2 to 4 months. On the other hand the feeding of methylcholanthrene or its injection into the gastric mucosa rarely results in an adenocarcinoma, and in such cases very long contact is required. The speed and sphere of action of the hydrocarbon is greatly enhanced in embryonic tissue. Thus fragments of embryonic mouse skin treated with the substance and transplanted to adult mice contain epidermoid carcinomas in from 30 to 40 days, and adenocarcinomas are induced in embryonic mouse stomachs with an incidence close to 100 per cent in a comparable length of time (Plate IV, Fig. 23-24).

The increased susceptibility of embryonic transplants to carcinogenic agents has been used experimentally to aid in the testing of suspected materials, and although the findings in animals do not necessarily apply to man, some of the results are comforting. Tobacco has been suggested as an etiological factor in the present high incidence of lung cancer. Accordingly fragments of embryonic lung were treated with various tobacco tars and extracts and transplanted to experimental animals. The transplants were histologically intact at the end of a 3-month period, and there were no changes to indicate that the tobacco exerted a carcinogenic effect.

Embryonic tissues are susceptible to other noxious agents. The typical lesions of asbestosis are reproduced in embryonic mouse lung infiltrated with asbestos fibers and transplanted to adult animals, and beryllium added to transplants of embryonic tissue induces lesions comparable to those found in man. Study of these diseases and a variety of others is facilitated by the rapidity of the tissue reaction and the wide selection of transplantation sites. The technique is also of value in the investigation of more general problems in growth and development. For example it has been suggested that the fetal adrenal may have a special function distinct from that of the adult organ and controlled by gonadotropic rather than adrenotropic hormone. The influence of the two hormones on the fetal adrenal can be readily followed in anterior chamber transplants, and it is of some interest that growth is accelerated by gonadotropic hormone and retarded by adrenotropic hormone. Again fragments of embryonic mouse heart muscle continue to beat after anterior chamber transfer, and the effect of various drugs on the rhythm or rate can be determined and measured by direct visual observation.

B. Heterologous transplantation. All of the various homologously transplantable embryonic tissues have also been successfully transferred to other species. The brain and the eye have proven to be the most efficient sites in all species, but the subcutaneous space of the hamster is also an excellent nidus for heterologous growth. There is a high incidence of takes and the transplants persist for long periods of time. Histologically they resemble adult organs in all particulars (Plate V, Fig. 25-27).

The function of heterologous transplants has been less extensively studied than that of homologous transplants, but it has been found that adrenalectomized guinea pigs can be maintained on transplants of embryonic rabbit adrenals and that transplants of rabbit ovaries growing in guinea pig eyes respond to gonadotropic hormone. The susceptibilities characteristic of embryonic tissues transplanted homologously are retained in heterologous transplants. Thus embryonic rabbit skin infected with the Shope papilloma virus responds with the formation of characteristic papillomas after transfer to alien species, and embryonic chicken skin infected with the Rous sarcoma virus or the fowl pox virus and transplanted to mouse brain develops the typical disease (Plate V, Fig. 28-30) (Albrink & Greene, 1953; Greene, 1953).

It should be emphasized that embryonic human organs and tissues as well as those of laboratory animals survive and grow on transfer to alien species (Plate VI, Fig. 31-36). In such locality normal human tissues are available for experimentation not permissible in the natural host. Several investigative approaches have been followed. For example there is a question relative to the carcinogenicity of methylcholanthrene for human tissues; the material is effective in producing tumors in mice and rats, but the tissues of rabbits are unresponsive. In our experiments embryonic human tissue was infiltrated with methylcholanthrene and transplanted to the eyes of guinea pigs. The animals were held under observation for 13 months, and although the transplanted tissue had persisted throughout the period, histological examination at its end showed no change suggestive of neoplasia.

III. HYPERPLASTIC AND PRECANCEROUS TISSUES

The term "benign tumor" is an ambiguous one and will not be employed in the present discussion. The term is used clinically in reference to two biologically different types of growth. One of these is nothing more than a localized area of hyperplasia made up of mature, adult, postmitotic cells with fully expressed potentialities. Such so-called "tumors" as the osteomas and many of the adenomas, particularly those of ductless glands, belong to this category. These growths, being composed of adult cells, show as might be expected exactly the same transplantation reactions as do adult tissues, that is they survive autologous and homologous transplantation but fail to grow on heterologous transfer.

Growths of the second type are of much more biological interest. They contain a greater complement of partially differentiated intermitotic cells whose potentialities have not reached full expression and which frequently develop into cancer. Such tumors as the breast papillomas and the rectal polyps belong to this category and show a characteristic set of transplantation reactions unlike those of any other tissue state.

A tumor such as a breast papilloma will grow when transplanted back elsewhere in the primary host but always fails to grow when transplanted to a normal unrelated animal. Despite its failure to survive in a normal

[183]

animal, the tumor will grow when transferred to another individual bearing a breast papilloma (Plate VII, Fig. 37-39). The occurrence of growth under such restricted circumstances is highly significant and the finding has been confirmed utilizing other tumors. Survival of the tumor at this stage of development appears to be dependent on the special constitutional status of the spontaneous tumor-bearing animal. The factors concerned in this status are not operative in normal animals and accordingly the tumors will not survive in normal animals.

Investigation has shown that the factors are present in tumor-bearing animals from the earliest stages of tumor development and are specific for different types, that is a breast papilloma will not grow in an animal bearing a developing epidermoid carcinoma but requires the special factors operative in the breast tumor animal. Some indication of the nature of these factors has been obtained from a continued study of tumor-bearing rabbits, particularly of a breast-tumor strain (Greene, 1939; 1940). Such animals show profound endocrinological changes in addition to the tumor, and the changes are comparable to those produced in rabbits by the long-continued administration of estrogenic hormone. This association suggested an experimental approach, and tumors found to be dependent by test in normal animals were transplanted to a group of rabbits subjected to prior treatment with estrogen. Takes occurred in all of the animals, and it was concluded that in the case of breast tumors the factors on which survival and growth depend are functions of a constitutional status sequential to the long-continued action of estrogenic hormone.

It should be pointed out that the breast tumors represent a special case and there is no evidence of the operation of estrogenic hormone in the development of other tumors. The nature of the factors essential to the development of other tumors is unknown, but the factors are constitutional in distribution and there are indications of an endocrinological character in other special cases. The prostatic tumors are an example in point and the effect of castration is highly suggestive that such tumors may be dependent on androgenic hormone.

In any case the altered constitutional status of tumor-bearing animals is readily demonstrable, for in addition to providing a suitable nidus for the growth of dependent tumors of the same type the animals show a changed susceptibility to the transfer of heterologous cancer. The Brown-Pearce rabbit cancer grows poorly, if at all, in normal C_3H mice, but when transplanted to C_3H mice bearing spontaneous tumors, takes invariably occur and growth is rapid. Similarly the Rous chicken sarcoma

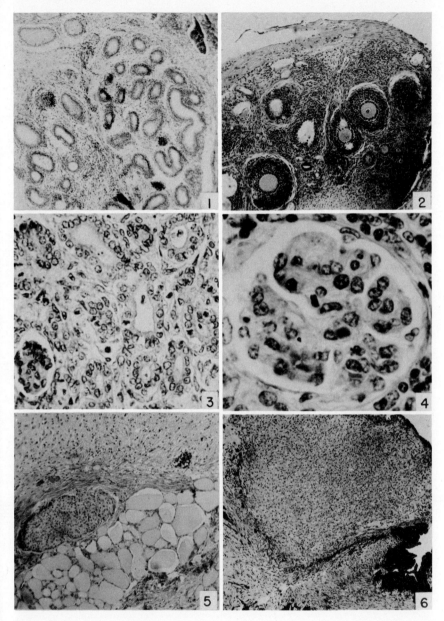

PLATE I.

Fig. 1. Transplant of adult rabbit testicle in anterior chamber of male rabbit killed 2 months after transfer. Mag. 85×.

Fig. 2. Transplant of adult guinea pig ovary in anterior chamber of male guinea pig killed 1.5 months after transfer. Mag. 110×.

Fig. 3. Transplant of adult guinea pig kidney in anterior chamber of guinea pig killed 1 month after transfer. Mag. 500×.

Fig. 4. Glomerulus from anterior chamber transplant of adult kidney 1.5 months after transfer. Capillaries contained red blood cells. Mag. 800×.

Fig. 5. Transplant of adult guinea pig thyroid and parathyroid glands in brain of guinea pig killed 3 months after transfer. Mag. 150×.

Fig. 6. Transplant of adult rabbit brain in anterior chamber of rabbit killed 1 month after transfer. Mag. 85×.

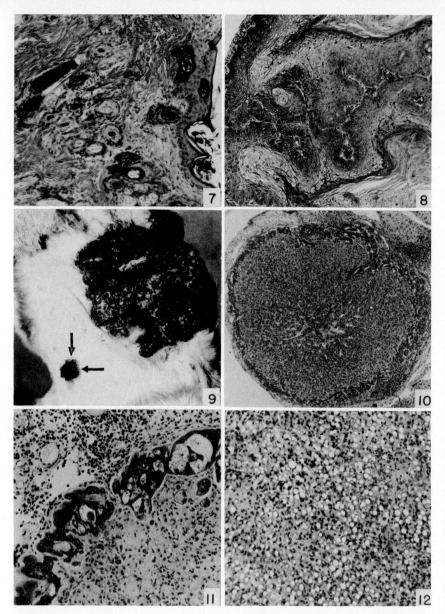

PLATE II.

Fig. 7. Transplant of adult rabbit skin in brain of rabbit killed 1.5 months after transfer. Mag. 185×.

Fig. 8. Papilloma arising in adult rabbit skin infected with the Shope virus and transplanted to a rabbit's brain. Section taken 1 month after transfer. Mag. 150×.

Fig. 9. Autologous transplant of skin infected with the Shope virus in immune rabbit. The large papilloma was induced by scarification of the skin and application of the Shope virus 4 months before the photograph was taken. Repeated attempts to reinfect the animal were unsuccessful but a typical papilloma developed in the autoplant (arrows). The skin of the autoplant was washed and infected with the virus *in vitro* before return to the animal.

Fig. 10. Transplant of embryonic guinea pig adrenal in eye of adult guinea pig. The adrenal was transferred in association with fragments of placenta. Note organization of gland into zones. The medulla never survives transfer. Section taken 2 months after transfer. Mag. 75×.

Fig. 11. Growing fragment of placenta from a different area (Fig. 10). Mag. 185×.

Fig. 12. Transplant of embryonic guinea pig liver to eye of adult pig. The liver was transferred in association with fragments of embryonic lung. The fatty change occurs in all liver transplants held longer than 3 weeks. Mag. 185×.

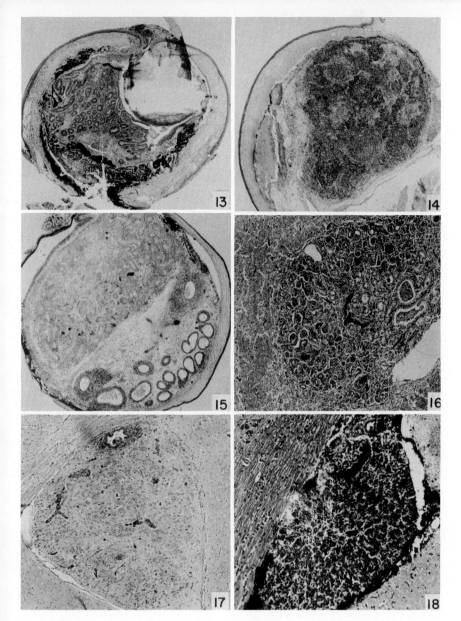

PLATE III.

Fig. 13. Transplant of embryonic mouse intestine in anterior chamber of adult mouse killed 2 months after transfer. Mag. 35×.

Fig. 14. Transplant of embryonic mouse spleen in eye of adult mouse killed 13 months after transfer. Mag. 70×.

Fig. 15. Transplant of embryonic mouse testicle and epididymis in eye of adult mouse killed 1 month after transfer. Mag. 30×.

Fig. 16. Transplant of embryonic guinea pig kidney in brain of adult guinea pig killed 1 month after transfer. Mag. 95×.

Fig. 17. Transplant of embryonic mouse brain to brain of adult mouse killed 1 month after transfer. Mag. 75×.

Fig. 18. Transplant of embryonic guinea pig pituitary to brain of adult guinea pig killed 2 months after transfer. Mag. 150×.

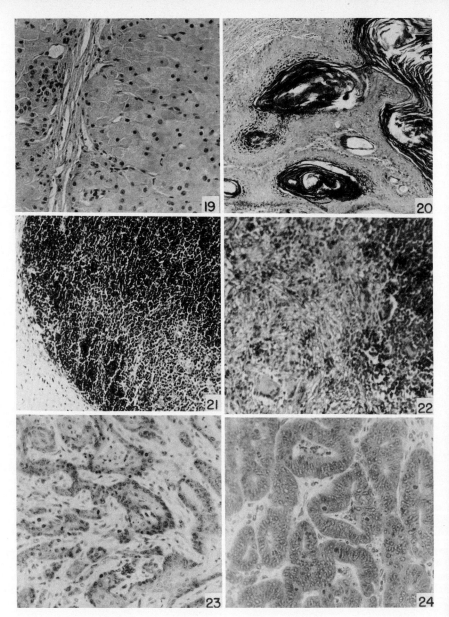

PLATE IV.

Fig. 19. Transplant of embryonic human adrenal in rectus muscle of patient with Addison's disease. Biopsy taken 8 months after transfer. Mag. 260×.

Fig. 20. Papilloma induced in embryonic rat skin treated with the Shope virus and transplanted into the brain of an adult rat. Section obtained 31 days after transfer. Mag. 140×.

Fig. 21. Transplant of embryonic mouse thymus in brain of adult mouse killed 2 months after transfer. Mag. 150×.

Fig. 22. Transplant of embryonic mouse thymus in brain of adult mouse. The embryonic organ was treated with a filtrate of human Boeck's sarcoid before transfer. Note granulomatous lesion characteristic of the human disease. Section taken 2 months after transfer. Compare with control transplant shown in Fig. 21. Mag. 250×.

Fig. 23. Transplant of embryonic mouse urinary bladder treated with methylcholanthrene in subcutaneous space of adult mouse. Note epidermoid carcinoma. Animal killed 40 days after transfer. Mag. 250×.

Fig. 24. Transplant of embryonic mouse colon treated with methylcholanthrene in subcutaneous space of adult mouse killed 43 days after transfer. Note adenocarcinoma. Mag. 250×.

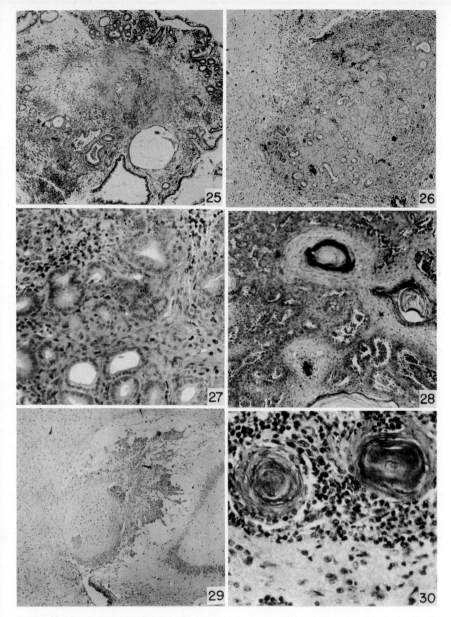

PLATE V.

Fig. 25. Transplant of embryonic **rabbit** stomach in eye of adult guinea pig killed 1 month after transfer. Mag. 50×.

Fig. 26. Transplant of embryonic guinea pig kidney to brain of adult mouse killed 1 month after transfer. Mag. 75×.

Fig. 27. Transplant of embryonic chick testicle in eye of adult guinea pig killed 4 months after transfer. Mag. 340×.

Fig. 28. Papilloma arising in transplant of embryonic rabbit skin treated with the Shope virus in the brain of an adult rat killed 1 month after transfer. Mag. 150×.

Fig. 29. Rous sarcoma arising in transplant of embryonic chick tissue treated with the Rous virus in the brain of an adult mouse killed 2 weeks after transfer. Mag. 75×.

Fig. 30. Fowl pox lesion in embryonic chick skin treated with fowl pox virus and transplanted to the brain of an adult mouse. Section taken 10 days after transfer. Mag. 400×.

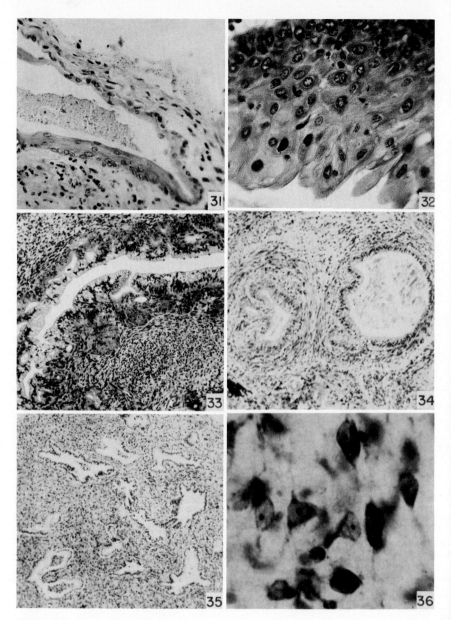

PLATE VI.

Fig. 31. Transplant of embryonic human skin in eye of adult guinea pig killed 3 months after transfer. Mag. 250×.

Fig. 32. Transplant of embryonic human skin in eye of adult guinea pig killed 1 month after transfer. Mag. 375×.

Fig. 33. Transplant of embryonic human stomach mucosa in eye of adult guinea pig killed 1 month after transfer. Mag. 150×.

Fig. 34. Transplant of embryonic human intestine in eye of adult guinea pig killed 1 month after transfer. Mag. 150×.

Fig. 35. Transplant of embryonic human lung in eye of adult rabbit killed 1 month after transfer. Mag. 130×.

Fig. 36. Transplant of embryonic human brain in eye of adult guinea pig killed 4.5 months after transfer. Note ganglion cells. Mag. 900×.

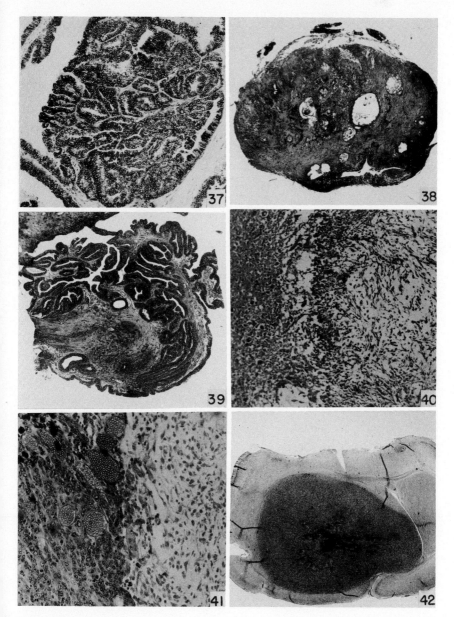

PLATE VII.

Fig. 37. Spontaneous papilloma in breast of rabbit. Fragments used for transfers illustrated in Figs. 38 and 39. Mag. 125×.

Fig. 38. Transplant of papilloma shown in Fig. 37 in eye of normal adult rabbit 1 month after transfer. No take. Mag. 75×.

Fig. 39. Transplant of papilloma shown in Fig. 37 in eye of another rabbit bearing a spontaneous breast papilloma. Take. Mag. 75×.

Fig. 40. Transplant of Rous chicken sarcoma in brain of adult guinea pig killed 12 days after transfer. Mag. 100×.

Fig. 41. Transplants of the Rous chicken sarcoma and a mouse neuroblastoma, C1300, growing together in a guinea pig eye. Section taken 15 days after transfer. Mag. 200×.

Fig. 42. Mouse brain bearing transplant of a human glioblastoma multiforme. Animal killed 60 days after transfer. Mag. 12.5×.

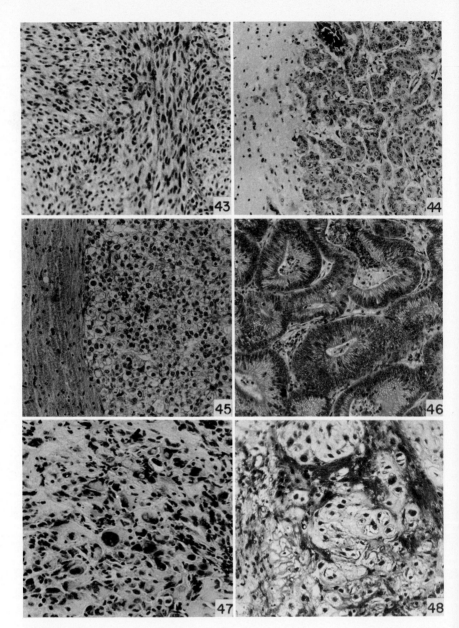

Fig. 43. Transplant of human glioblastoma multiforme in brain of guinea pig killed 2 months after transfer. The transplant reveals the neoplastic element of the tumor to be a spongioblast. Mag. 280×.

Fig. 44. Transplant of human colonic carcinoma in brain of a guinea pig killed 57 days after transfer. Mag. 200×.

Fig. 45. Transplant of human mammary carcinoma in brain of a guinea pig killed 60 days after transfer. Mag. 190×.

Fig. 46. Transplant of human rectal carcinoma in eye of a guinea pig killed 4 months after transfer. Mag. 200×.

Fig. 47. Transplant of human fibrosarcoma in eye of guinea pig killed 40 days after transfer. Mag. 275×.

Fig. 48. Transplant of human chondrosarcoma in eye of guinea pig killed 54 days after transfer. Mag. 230×.

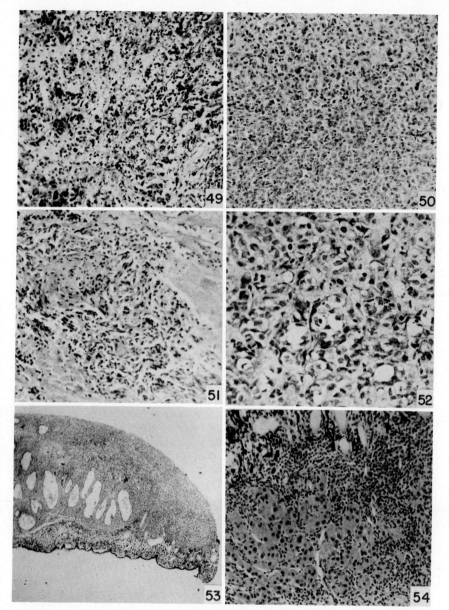

PLATE IX.

Fig. 49. Section of enlarged lymph node from groin of middle-aged woman showing necrosis, desmoplasia, and anaplastic tumor cells. Histological picture does not allow identification of tumor. Mag. 200×.

Fig. 50. Transplant of human tumor shown in Fig. 49 in guinea pig eye. Histology allows identification of tumor as a malignant melanoma. Mag. 200×.

Fig. 51. Section of tumor from vertebral column of young man. Identification not possible. Mag. 165×.

Fig. 52. Transplant of tumor shown in Fig. 51 in guinea pig eye. Histology allows identification of tumor as a chordoma. Mag. 200×.

Fig. 53. Low magnification of transplant of human hypernephroma and embryonic guinea pig lung in eye of adult guinea pig. Mag. 30×.

Fig. 54. Higher magnification of section of transplant shown in Fig. 53. Note growth of hypernephroma. Mag. 185×.

grows on subcutaneous transfer to tumor-bearing C_3H mice, but takes have not been obtained in normal C_3H mice.

Whatever its nature the altered constitutional status of the tumor-bearing animal provides factors essential for the continued growth and development of the tumor. Such factors are not supplied by normal animals and the tumors can not be transplanted to normal animals (Greene, 1951).

IV. CANCER TISSUE

With continued residence in the primary host or in other tumor-bearing animals, the neoplastic focus such as the breast papilloma cited above becomes independent of the factors concerned in growth and development and hitherto essential to its existence. At this stage of development the growth is referred to as a cancer and its autonomy or independence of conditioning factors is such that it will survive transplantation to normal animals irrespective of species (see numerous papers by Greene from 1938 to 1954). The attainment of autonomy is a late development in the history of a tumor and is associated with the occurrence of metastasis and a rapid acceleration in the fatal course of the disease.

Cancer tissue like embryonic tissue is transplantable autologously, homologously, and heterologously. Its transfer to an alien species is not associated with a foreign body reaction, growth is rapid, and there is no evidence of tissue incompatibility. On the contrary one of the most impressive biological characteristics of cancer is its wide compatibility and the absence of features of sufficient specificity to elicit reactions. In fact the differences between zoological classes do not constitute barriers to the transfer of cancer, and the Rous chicken sarcoma grows readily in rabbits, guinea pigs, and mice (Plate VII, Figs. 40-41) (Albrink and Greene, 1953; Shrigley, Greene and Duran-Reynals, 1945; 1947). Further, immunization of guinea pigs against human tissues does not interfere with the transplantation and growth of human cancer.

The development of autonomy is not necessarily associated with a morphological change in the tumor but is coincident with the attainment of the ability to metastasize. A determination of heterotransplantability is thus of prognostic value in dealing with human cancer (Greene, 1952). In a recent series of 123 human cancers derived from the operating room, 65 proved to be transplantable and 58 failed to grow. At the present time the mortality in the heterotransplantable group has been 93.8 per cent in contrast to 20.7 per cent in the nonheterotransplantable group. The point is further emphasized by a comparison of the mortality in the two groups at 6-month intervals after operation. During the first 6 months after operation, the mortality in the heterotransplantable group was 63 per cent and 1.7 per cent in the nonheterotransplantable group. At 12 months the contrasting figures were 84.6 per cent and 3.4 per cent; at 18 months, 95.2 per cent and 5.8 per cent; and at 24 months, 98.3 per cent and 6.9 per cent respectively. None of the patients in the hetero-transplantable group survived more than 30 months whereas the mortality in the nonheterotransplantable group 30 months after operation was only 13.8 per cent.

The growing transplants closely resemble the human cancer in the majority of cases, but occasionally extremely anaplastic tumors show a higher degree of differentiation and organization in the guinea pig than in the biopsy specimen. In such cases guinea pig transfer allows a more precise classification of the tumor and sometimes suggests a different primary site (Plate VII-IX, Fig. 42-52).

The point of primary interest in the present discussion is the profound biological transformation characterizing the development of cancer. The biological properties of cancer are not sudden mutations in normal cells but on the contrary represent the final step in a developmental process. The dependency of the primary neoplastic focus restricts its transplanta-bility to highly individualized animals, but with continued development in such an environment the focus eventually attains independence of the factors supplying its special requirements and the resulting autonomy transcends species barriers.

Special cases exist in which the factors pertinent to dependency reside in the tumor cell rather than in the constitution of the host. Such tumors are transplantable from inception, for the conditioning factor necessarily accompanies the cell on transfer, and in such cases heterotrans-plantability is not an expression or autonomy. The Shope rabbit papilloma and the Rous chicken sarcoma are outstanding examples (Greene, 1953). Both of these tumors are transplantable from inception for the condition-ing factor is an intracellular virus and by virtue of its location is trans-ferred to the new host along with the tumor cell.

The virus induced tumors are thus independent of the constitution of the host in their transplantability and will grow in heterologous as well as homologous species. With continued development the virus-induced tumors attain independence of their conditioning factors (Syverton et al., 1950). The epidermoid carcinoma arising from a Shope papilloma is independent of the virus concerned in the growth and evolution of the papilloma, and the conversion is accompanied by disappearance of the virus. The development of independence of the virus, or autonomy, is associated with the attainment of the ability to metastasize, and this sequence is again a reflection of the behavior of other tumors.

A further paradox in the relationship between heterotransplantability and other manifestations of autonomy is brought out by the behavior of human brain tumors. Both of the two more malignant brain tumors, the glioblastoma multiforme and the medulloblastoma, are heterotransplantable, yet neither metastasize. The usual explanation of the failure to metastasize involves the existence of a hypothetical blood-brain barrier preventing entrance of tumor cells to the blood stream. However, tumors originating in non-nervous tissue do metastasize when transplanted to the brain. For this reason it seems more probable that the failure to metastasize relates to the character of the tumor cell rather than to its site of growth, and investigations to be discussed in succeeding paragraphs suggest that the pertinent factor is an inability to elicit stroma and vascular supply in other bodily regions.

V. TRANSPLANTATION SITE

The results of the various transplantation experiments are arranged in Table I to compare the reactions of different tissue states in animals of different constitutional type, and it is apparent that the success of

TABLE I. Transplantation reactions of various tissue states in animals of different constitutional types.

Tissue state	TYPE OF TRANSFER		
	Autologous	Homologous	Heterologous
Adult	+	+	—
Hyperplastic	+	+	—
Embryonic	+	+	+
Precancerous	+	—	—
Cancerous	+	+	+

+ indicates growth
— indicates no growth

transfer varies in relation to the status of the transplant whether adult, embryonic, precancerous, or cancerous. This summary represents a generalization based on optimal conditions and fails to show the existence of two additional variables—one pertaining to qualities of the transplant distinct from tissue state and the other to the site of transplantation. A consideration of these variables is important, for in many instances their operation determines the success or failure of transfer.

In the majority of cases the transfer of a fragment of tissue from one individual to another is associated with death of the stromal component of the graft, and the parenchyma persists only if a connective tissue scaffolding and a vascular supply are furnished by the new host. The reaction of the new host to the graft in this respect is determined by the stroma-evoking properties of the parenchymal cells on the one hand and by the responsiveness of the connective tissue at the transplantation site on the other.

The ability of transplanted cells to elicit stroma varies within wide limits. The epithelial cells of the kidney or breast are potent stroma inducers, whereas cells derived from the adrenal or brain have minimal capacities in this direction. In general there appears to be a direct relationship between stroma-inducing ability and the normal stromal complement of the organ from which the cells are derived. An abundance of stroma is essential to the architectural pattern of the kidney and breast and the cells of these organs possess strong stroma-inducing properties. On the other hand such organs as the adrenal and brain contain little stroma and their cells in turn prove to be exceptionally poor stroma evokers.

The connective tissues in different regions of the body also vary in their responsiveness to the same stroma-inducing stimulus. The connective tissue of the subcutaneous space in most animal species is least responsive while that of the iris and brain is highly reactive. A poor stroma-inducer such as the adrenal gland usually fails to elicit stroma from the unresponsive connective tissue of the subcutaneous space and the transplant dies. On the other hand the minimal stimulus is sufficient to induce an adequate stroma from the more responsive tissues of the iris or brain, and a take occurs.

The situation with respect to the brain is provident. Brain parenchyma is an extraordinarily poor stroma inducer, yet its connective tissue, although present in minimal amounts, is the most responsive of the body. If these relationships were altered in the direction of an increase in stroma-inducing capacity, the resulting fibrosis would still further increase the difficulties involved in dealing with hospital administrative officers.

Brain parenchyma is capable of eliciting stroma from the eye or the brain itself, but no stromal response occurs in the subcutaneous space, and despite innumerable attempts it has not been possible to transplant brain to any bodily region other than the eye and brain. On the other hand tissues incapable of eliciting a stromal response in other bodily regions are readily stromatized on transfer to the brain. The hypernephroma of the human kidney is a tissue in point. This tumor, like other tissues with a scanty connective tissue content, is a poor stroma inducer. In the primary host it may invade and extend along the renal vein for some distance, but despite the fact that cells must inevitably be washed into the circulation, distant foci of growth may not be found. On transfer to the guinea pig eye, its stroma-inducing properties are too slight to elicit a stroma from the highly sensitive connective tissue of the iris and its growth resembles that of a tissue culture. However, the connective tissue of the brain responds to the same stimulus with an adequate proliferation, and a well-stromatized transplant results.

It seems probable that a similar situation applies with respect to the human brain tumors that grow on heterologous transfer but fail to metastasize. In essence the tumors represent anaplastic forms of brain cells and as such share or possess, in even diminished order, the poor stroma-inducing qualities of brain cells. It is suggested that these qualities are sufficient to evoke stroma in the brain but insufficient to stimulate a response of connective tissue in other regions adequate to support and nourish the disseminated tumor cells. The results of transplantation experiments are in line with this suggestion, for all attempts to transfer the tumors to sites other than the brain and eye have been unsuccessful and a variety of techniques have been employed.

One of the techniques unsuccessfully employed in the case of brain or brain tumors, but highly effective in other instances, depends on the fact that the connective tissue contained in embryonic organs possesses a remarkable degree of reactivity and that the embryonic tissue is also an excellent stroma-inducer. The procedure evolved represents an attempt to utilize their potentiality for the production of stroma with the embryonic tissue, the middleman in the exchange. In practice a small fragment of embryonic tissue is added to the inoculum of tissue known to be a poor stroma-inducer, such as a human hypernephroma, and the two fragments are transplanted simultaneously to the new host. The embryonic portion of the transplant elicits a stroma at the new site and in turn provides the hypernephroma with a connective tissue scaffolding and blood supply (Plate IX, Fig. 53-54). The use of this technique has made it possible to transfer many tumors and tissues to regions where under normal conditions no growth could be obtained.

It is important to emphasize that tumor stroma is not a neoplastic element and does not share the biochemical and immunological attributes that distinguish the parenchyma. However, stroma and parenchyma are so intimately associated anatomically that they can not be separated for transplantation purposes. Thus occasions occur in which the stroma present in a transplant is sufficiently different from the tissues of the new host as to evoke a foreign-body reaction, and the transplant fails despite the fact that the parenchyma or essential neoplastic element of the tumor and the new host are entirely "compatible." For example in testing the transplantability range of a tumor from a C_3H mouse, it was found that transfer to C57 blacks, dba's and several other varieties of mice resulted in takes in all strains except the C57 black. However, subsequent transfer of the tumor growing in dba's to C57 blacks resulted in growth. The parenchyma was identical in the two tumors and they differed only in stromal content, the stroma of one tumor being made up of C_3H connective tissue and the other of dba connective tissue. Apparently the C_3H connective tissue was sufficiently different from the tissues of the C57 mouse to induce a reaction inhibiting growth whereas the dba connective tissue, which replaced the C_3H connective tissue after transfer to dba mice, was "compatible" with the tissues of the C57.

VI. COMPATIBILITY AND NONCOMPATIBILITY

In view of the results described it is suggested that the terms "compatibility" and "noncompatibility" as applied to tissue transplantation require qualification in relation to factors pertaining to the transplant and the transplantation site as well as to the constitution of the donor and the recipient. The general statement, that the tissues of a certain species, strain, or individual are incompatible with those of another, disregards both the status of the tissue and the site of transplantation and derives largely from considerations based on adult tissues as alone representative of the donor and on the subcutaneous space as alone representative of the recipient. Actually the status of a tissue varies with development and may be embryonic or cancerous as well as adult. Further, the body of the recipient, extending beneath the subcutaneous space, contains areas of differentiation such as the brain and eye, whose special attributes provide the tissue with more suitable conditions for growth but do not confer an independence of the constitutional factors determining the reaction of transplantation.

The point to be stressed is that the eye and the brain are parts of the body and are not unconstitutional or subversive in their response to transplantation. Moreover the adult state of a tissue represents a developmental phase and not its life history; the initial state of all tissues is embryonic and a terminal state may be cancerous. Compatibility relationships vary accordingly. On a basis of anterior chamber reactions, the parenchyma of the human brain in embryonic and cancerous states is compatible with the guinea pig while in an adult state the tissue is not compatible with such a constitution. On a basis of reactions in the subcutaneous space on the other hand, all developmental states of the human brain are incompatible with the constitution of the guinea pig. Investigation of this anomaly suggests that the discrepancy arises from local causes which operate in the subcutaneous space to obscure the expression of constitutional factors.

The brain is an extreme example for illustration, but the reaction to other embryonic tissues is similar with variations corresponding to their stroma-inducing capacities. The fact that failure to grow in the subcutaneous space relates to deficiencies in this capacity rather than to incompatibility can be demonstrated in certain cases by the coincident transfer of a second tissue which acts both as an auxiliary stroma inducer and as a stroma provider. Other regional influences operating to differentiate the reaction to transplantation in the subcutaneous space and the eye are apparent from the results of tumor transfer. The Brown-Pearce and V-2 rabbit carcinomas grow on transfer from the rabbit to the subcutaneous space of hamsters, rats, and mice, but in the majority of other cases immediate heterologous transfer to this site is unsuccessful. However, if the tumor is first grown in the eye or brain of the new species, subsequent transfer to the subcutaneous space or another region of the body results in takes. In such cases the initial failure can not be interpreted to mean that the tissues of the tumor and of the new host are incompatible. It would seem rather to result from some adverse reaction between the stromal component of the tumor and the subcutaneous tissues of the heterologous animal, for when the original stroma of the tumor is replaced by the connective tissue of the new species as a consequence of anterior chamber growth, such a reaction does not occur.

The failure of early premetastasizable tumors to grow on simple homologous transfer appears related to an inability of the normal animal to provide some element essential to the growth of the tissue rather than to an incompatibility factor. Transfer to animals of similar genetic composition but known to carry the essential element results in growth, and in some instances the missing element can be added to the normal animal. Further, an interpretation of the failure to grow as a manifestation of incompatibility requires the assumption that compatibility is an ac-

[191]

quired rather than a genetic character, for with continued development the tumors do attain the ability to grow in normal animals.

The inability of normal adult tissues to survive heterologous transfer may be taken as a clear-cut instance of incompatibility, but even here the issue is not clear and the term fails to satisfy experimental observations. A point with reference to heterologous embryonic transplants of particular concern is that, although the transplants differentiate and mature into adult tissues, they do not elicit the response that characterizes the transplantation of adult tissues of the foreign species. The transfer of adult rabbit skin to the mouse results in an immediate foreign-body reaction, but the maturation of embryonic rabbit skin to adult rabbit skin in the mouse is not accompanied by any sign of inflammatory disturbance. The antigenic qualities incident to the "rabbitness" of the transplanted adult tissue are supposedly responsible for the foreign body reaction, yet in view of its susceptibility to the Shope papilloma virus the maturing embryonic transplant is also rabbit in nature. A number of unexplored possibilities suggest themselves. It is conceivable that the transplant is not rabbit in nature and that susceptibility to the virus relates to some quality other than "rabbitness." Such a possibility is given support by the demonstrated susceptibility of embryonic rat tissue as noted above. On the other hand it is also conceivable that the species factors determining susceptibility are different from those determining antigenicity and that expression of the latter factors depends on the nurture rather than the nature of the tissue. Accordingly embryonic rabbit tissue developing to maturity in the rabbit possesses the antigenic attributes of the species whereas embryonic rabbit tissue developing to maturity in the mouse fails to evolve such qualities. The inferences attached to such speculations are of interest from several points of view and form the basis of continued investigations.

BIBLIOGRAPHY

Albrink, W. S., and H. S. N. Greene. 1953. The transplantation of tissues between zoological classes. *Cancer Research 13*, 64.

Greene, H. S. N. 1938. Heterologous transplantation of human and other mammalian tumors. *Science 88*, 357.

Greene, H. S. N. 1939. Familial mammary tumors in the rabbit. I. Clinical history. II. Gross and microscopic pathology. III. Factors concerned in their genesis and development. *J. Exp. Med. 70*, 147.

Greene, H. S. N. 1940. Familial mammary tumors in the rabbit. IV. The evolution of autonomy in the course of tumor development as indicated by transplantation experiments. *J. Exp. Med. 71*, 305.

Greene, H. S. N. 1941. Heterologous transplantation of mammalian tumors. I. The transfer of rabbit tumors to alien species. *J. Exp. Med. 73*, 461.

Greene, H. S. N. 1941. Heterologous transplantation of mammalian tumors. II. The transfer of human tumors to alien species. *J. Exp. Med. 73*, 475.

Greene, H. S. N. 1942. Heterologous transplantation of a human fibrosarcoma. *Cancer Research 2,* 649.

Greene, H. S. N. 1943. The heterologous transplantation of embryonic mammalian tissues. *Cancer Research 3,* 809.

Greene, H. S. N., and P. K. Lund. 1944. The heterologous transplantation of human cancers. *Cancer Research 4,* 352.

Greene, H. S. N., and E. D. Murphy. 1945. The heterologous transplantation of mouse and rat tumors. *Cancer Research 5,* 269.

Greene, H. S. N. 1945. The production of carcinoma and sarcoma in transplanted embryonic tissues. *Science 101,* 644.

Greene, H. S. N. 1946. The heterologous transplantation of mouse tumors induced *in vitro. Cancer Research 6,* 306.

Greene, H. S. N. 1946. The microscope or the guinea pig? *Yale J. Biol. Med. 18, 239.*

Greene, H. S. N. 1949. Heterologous transplantation of the Brown-Pearce Tumor. *Cancer Research 9,* 728.

Greene, H. S. N. 1950. The heterologous transplantation of human melanomas. *Yale J. Biol. Med. 22,* 611.

Greene, H. S. N. 1951. The transplantation of tumors to the brains of heterologous species. *Cancer Research 11,* 529.

Greene, H. S. N. 1951. A conception of tumor autonomy based on transplantation studies: a review. *Cancer Research 11,* 899.

Greene, H. S. N. 1952. The significance of the heterologous transplantability of human cancer. *Cancer 5,* 24.

Greene, H. S. N. 1953. The induction of the Shope papilloma in transplants of embryonic rabbit skin. *Cancer Research 13,* 58.

Greene, H. S. N. 1953. The transplantation of human brain tumors to the brains of laboratory animals. *Cancer Research 13,* 422.

Greene, H. S. N. 1953. The heterologous transplantation of human lung cancer. *Cancer Research 13,* 347.

Greene, H. S. N. 1953. The induction of the Shope papilloma in homologous transplants of embryonic rat skin. *Cancer Research 13,* 681.

Greene, H. S. N. 1953. Transplantation of the Shope papilloma and the Rous sarcoma during early developmental stages. *Cancer Research 13,* 726.

Greene, H. S. N., and H. Arnold. 1945. The homologous and heterologous transplantation of brain and brain tumors. *J. Neurosurg. 2,* 315.

Shrigley, E. W., H. S. N. Greene, and F. Duran-Reynals. 1945. Studies on the variation of the Rous sarcoma virus following growth of the tumor in the anterior chamber of the guinea pig eye. *Cancer Research 5,* 356.

Shrigley, E. W., H. S. N. Greene, and F. Duran-Reynals. 1947. Growth of avian tumors other than the Rous sarcoma in the anterior chamber of the guinea pig eye. *Cancer Research 7,* 15.

Syverton, J. T., H. E. Dascomb, E. B. Wells, J. Koomen, and G. P. Berry. 1950. The virus-induced rabbit papilloma-to-carcinoma sequence. II. Carcinomas in the natural host, the cottontail rabbit. *Cancer Research 10,* 440.

Syverton, J. T., E. B. Wells, J. Koomen, H. E. Dascomb, and G. P. Berry. 1950. The virus-induced rabbit papilloma-to-carcinoma sequence. IV. Immunological tests for papilloma virus in cottontail carcinomas. *Cancer Research 10,* 474.

X. SPECIFICITY IN GROWTH CONTROL

BY PAUL WEISS[1]

IN ESSENCE, this chapter is a sequel to the general discussion of "Specificity in Development and Growth" given before the Society for the Study of Development and Growth in 1945 (P. Weiss, '47). Of the problems raised then, that of growth control will be singled out here for special consideration. The account will be confined to illustrative examples, with no attempt at reviewing the subject comprehensively.

"Specificity" is understood here as that property of two interacting systems, *A* and *B*, which permits *A* to react to *B* with some degree of selectivity. Biological specificity, as previously outlined, is a basic property of living systems and is most prominently displayed in drug responses, hormone actions, gene effects, immunological reactions, host-parasite relations, and developmental mechanisms. A common denominator in biochemical, presumably stereochemical, terms is indicated. Translated to these terms, specificity in growth confronts us with four separate issues: (1) the fact that the various cell strains of the body are different biochemically; (2) the possibility that biochemical distinctiveness may be not only a by-product of differentiation but also an instrumental factor in growth; (3) the possibility that this may constitute a mechanism for the humoral coordination and regulation of growth processes throughout the organism; and (4) the hypothesis that this regulatory function presupposes the generation in each cell strain of paired compounds of complementary configuration, after the antigen-antibody scheme.

I. THE BIOCHEMICAL MOSAIC

It has been stressed on previous occasions (Weiss, '49) that cells are "speciated" into biochemically diverse strains, indeed strains of much greater diversity and subtlety than are morphologically discernible. My attention was first drawn to this fact by the discovery of the

[1] Department of Zoology, University of Chicago. Present address of the author is: Rockefeller Institute for Medical Research, New York, N.Y. This chapter is dedicated by the author to Professor F. Baltzer of the University of Bern, Switzerland, on the occasion of his seventieth birthday. It represents a condensed version of the paper of the same title presented at the 1953 Growth Symposium.

Original work referred to in this chapter has been aided by grants from the Wallace C. and Clara A. Abbott Memorial Fund, University of Chicago; the American Cancer Society upon recommendation of the Committee on Growth of the National Research Council; and the National Institutes of Health, Public Health Service.

strict constitutional specificity of individual muscles and skin territories (Weiss, '52), but all the localized effects of hormones and drugs on different predisposed cell groups point to the same conclusion. Different organs are thus composed of cell strains of specifically different biochemical constitutions. On the other hand, organs occurring in pairs or other multiples may be assumed to be composed of biochemically similar cells. Such similarity of composition might provide a means of selective chemical communication between homologous cell groups regardless of spatial separation.

The existence of some such active interrelation between paired structures is indicated, for instance, by the observation that injury to a given nerve is often followed by an involvement of the symmetrical nerve (Greenman, '13, Nittono, '23, Koester, '03, Tamaki, '36). It is more definitely evinced by the compensatory growth reactions of one of a pair of organs after removal of the other.

II. HOMOLOGOUS COMPENSATION

A "spontaneous" spurt of growth of the residual parts of a partially removed organ system has been described for many objects: contralateral appendages in annelids (Zeleny, '02, '05); claws in Crustacea (Przibram, '07); kidneys in mammals (Golgi, 1882, Ribbert, '04, Arataki, '26, Rollason, '49); testes in mammals (Ribbert, 1895) and fishes (Robertson, '54); lungs (Haasler, 1891); and orbital glands (Teir, '51). Liver regeneration following partial ablation (e.g. Brues, '36) and blood cell regeneration after hemorrhage are in the same category. In all these cases the response is essentially confined to the homologous tissue and consists primarily of intensified reproduction of homologous protoplasm, which may take the form of cellular hypertrophy, hyperplasia, or regeneration. Necrosis may produce similar effects as extirpation.

The explanation of such compensatory growth has often been sought in the excessive functional load placed upon the remaining fraction of the organ system. However, neither the invertebrate cases nor the testes and orbital glands of the above list fit such an explanation. I, therefore, considered it more likely that these phenomena are manifestations of a much more general principle, namely the active maintenance of the total mass of each organ system in an equilibrium state and the return to that state after disturbance by virtue of a chemical communication system in which specific releases from each cell type, circulating in the body, would inform the homologous cell types of the state of their total mass. A test of this theory requires the demonstration that, (1) compounds produced by a given cell type have some selective effect on the same cell type, and (2) that this homologous effect is instrumental in the regulation of growth.

III. HOMOLOGOUS ORGAN-SPECIFIC EFFECTS

Evidence for the first point is contained in the observations of Danchakoff ('16) and of Willier ('24), who found enlarged spleens in chick embryos whose chorio-allantoic membranes had received spleen grafts. Weiss and Wang ('41), unaware of these results but corroborating them fully (see Weiss, '47), found that minute fragments of liver incorporated in the extraembryonic area of a chick embryo caused the host liver to grow to excessive dimensions. Implants of other tissues either had no effects or their effects were much less marked. The results with spleen were later confirmed and expanded by Ebert ('51). Along the same line, balancers implanted into the body cavity of urodele larvae affect the resorption of the host balancers specifically (Kollros, '40), as shown by the absence of a similar interference from implanted gills. Thus, there are definitely specific chemical effects transmitted humorally from a given type of organ to other cell groups of the same type, and these effects entail alterations of growth and size. Contrary to the inference from compensatory growth after partial removal, however, the experimental addition of tissue in these experiments produced no decrease, but rather a further increase of the homologous host tissue. Therefore, while the specificity of the growth reaction was clearly demonstrated, the sign of the reaction was paradoxical. This in itself was a significant revelation and we shall return to it below.

IV. IMMUNOLOGICAL REVERSAL OF ORGAN-SPECIFIC
COMPOUNDS

My original concept had been that: (1) cell populations of a given type keep their total mass in check by producing, as they grow, compounds which would repress the growth process in proportion to their concentration ("feed-back" fashion); (2) this self-inhibition would be due to steric complementariness between these inhibitor compounds and the specific catalysts of growth in each cell type; and (3) the assumed complementariness might be of the antigen-antibody type. On this assumption a series of experiments was started in 1938 to test the effects that antibodies to organ-specific compounds, rather than the compounds themselves, might exert on the embryonic growth of the homologous organs (Weiss, '39; see Weiss, '47).

The fact that antibodies to organ extracts may affect corresponding

[197]

organs selectively, mostly by damaging homologous cells, has been clearly demonstrated for the lens (Guyer and Smith, '18), the kidney (Smadel, '36, Pressman, '49), and others. Our injections of anti-liver and anti-kidney sera into chick embryos, however, entailed again larger sizes, rather than damage, of the homologous host organs. This could of course be ascribed to an initial damaging effect followed by repair with overcompensation. At any rate the specificity of the effect, if not its sign, seemed to have been ascertained, and the demonstration that organ-specific compounds did not lose their specific effectiveness by immunological (steric?) reversal seemed to give validity to the concept that these experiments were intended to test.

Although war work had interrupted this investigative program, its main premises and theoretical foundations were summarized in my discussion of specificity at the Growth Symposium in 1945. Purely as a guide in planning further research and in trying to reconcile the paradoxical results of the past, I have adhered to the particular concept of growth control that I had outlined previously (Weiss, '47, p. 272-273; '49, p. 180-181). Its pragmatic value has proven itself by bringing disparate results to a common denominator and by suggesting the design of new experiments reported further below.

V. A CONCEPT OF SPECIFIC GROWTH CONTROL

This concept is based on the following suppositions.

(1) Each specific cell type reproduces its protoplasm, i.e. "grows," by a mechanism in which key compounds that are characteristic of the individual cell type act as catalysts. The postulated cell-specific diversity of compounds is the chemical correlate of the "differentiation" of cell strains. Growth rate is proportional to the concentration of these intracellular specific catalysts (or "templates") in the free or active state. Under normal conditions these compounds remain confined within the cell.

(2) Each cell also produces compounds ("antitemplates") which can inhibit the former species by combining with them into inactive complexes. These may be turned out as direct by-products in the process of protoplasmic reproduction or be secondary differentiation products. They may be steric complements to the former or matched to them in some other fashion. The only prerequisites are: (a) that, contrary to the specific templates, they are released from the cell and get into the extracellular space and into circulation; (b) that they carry the specific tag of their producer cell type which endows them with selective affinity

for any cell of the same type; and (c) that they are in constant production so as to make up for their extracellular decomposition and final excretion.

(3) As the concentration of "antitemplates" in the extracellular medium increases, their intracellular density, hence inactivation of corresponding "templates," will likewise increase; in short, growth rate will decline in all cells belonging to that particular strain bathed by the common humoral pool. When stationary equilibrium between intracellular and extracellular concentration is reached, growth will cease. This mechanism results in a sigmoid growth curve for the total mass of each organ system (see Morales and Kreutzer, '45), and the familiar sigmoid curve for the whole organism would essentially be an aggregate of similar curves for the individual constituent organ systems.

This general concept offers a rational explanation for both the self-limiting character of growth in a confined medium (organism or culture) and the homologous organ-specific growth reactions after experimental interference. As can readily be seen, each interference will have to be examined in a dual light as to its effects on the concentration of both "templates" and "antitemplates," since it is the ratio of both that determines growth rate. The following conclusions can immediately be deduced from this scheme.

(a) Removal of part of an organ system removes part of the sources of the corresponding types of "templates" and "antitemplates." Since the former, according to our premise (1), have been in intracellular confinement, neither their former presence nor their recent loss are perceptible to other cells of the system. This is not so for the "antitemplates," which are in circulation and a reduction of whose production source would promptly be recognized by their lowered concentration in the extracellular pool. According to points (2) and (3) this would shift the intracellular ratio of "templates" to "antitemplates" temporarily in favor of the former, causing automatic resumption of growth till a steady state is restored—to all intents a "compensatory" growth reaction.

(b) Addition of a part should have opposite effects depending on whether or not its cells survive, or rather on the ratio of surviving to disintegrating cells. If all cells survive, the net effect would be an increased concentration in the circulation of the particular "antitemplates," hence a reduction in growth rate of the corresponding host system, provided it is still in a phase of growth (actual regression after growth has ceased need not be expected). On the other hand, cells that disintegrate release into the extracellular space a complement of specific "templates" that would otherwise never have escaped. Assuming that these, according to point (2), combine with or otherwise trap homologous "antitemplates," their presence in the pool will entail a temporary lowering of "antitemplate" concentration—hence again a spurt of growth in the homologous cell strains of the host. The simultaneous release of "antitemplates" from the disintegrating cells would have to be assumed to be insufficient to cancel this effect because of their faster metabolic degradation (see point 2c). An alternative possibility is that "templates" freed from cracked cells are directly adopted by homologous cells, where they would temporarily increase the intracellular concentration of growth catalysts—hence growth rate. In either scheme the release of cell content would accelerate homologous growth by increasing the intracellular ratio of "templates" to "antitemplates"—in the former case by reducing the denominator, in the latter case by increasing the numerator. It can be seen that in terms of this interpretation partial necrosis of an organ will have the same effect as partial removal, and that implantation of a fragment followed by some degeneration, as well as the injection of cell debris, are merely further variants of the same procedure.

To test the validity of this concept the following series of experiments were undertaken. In contrast to our earlier attempts, the assay of growth responses by measurements of size attained after a given period was abandoned as too unreliable; not only do such measurements fail to distinguish between the specific components of an organ and its content of connective tissue and blood, but also initial growth reactions can easily be missed due to secondary regulations or even overcompensations. Instead the mitotic index was introduced as a more sensitive and reliable criterion. A large series of experiments on the stimulation of mitotic activity in amphibian skin (Weiss and Overton; mostly still unpublished) has conclusively shown that cell division is secondary to cell growth; hence, if present it can be used as an index of protoplasmic increase. The same conclusion can be drawn from the precession of the increase of mass over that of cell number in liver regeneration (Brues, Drury, and Brues, '36).

VI. COMPENSATORY HYPERPLASIA WITHOUT FUNCTIONAL OVERLOAD

One of the inferences from our theory is that the "compensatory" growth of one member of a pair of organs after the removal or destruction of the other would be attributable to the disturbance of the described chemical equilibrium rather than to the burden of augmented functional activity. In view of the well-established compensatory hypertrophy of the remaining kidney after unilateral kidney removal, this experiment was repeated with the embryonic metanephros of the chick at a stage prior to the onset of its excretory function, the latter function being still fully exercised by the mesonephros. Wayne Ferris in our laboratory cauterized one metanephros in 12- to 13-day embryos and counted the mitotic response in the undamaged residual kidney fixed within 2 days after the operation. On the basis of a total count of 12,000 mitoses in 12 controls and 15 experimental cases, 4 of them sham operations, an average increase of 70 per cent was noted on the unharmed side that could be definitely identified as a response to the destruction of kidney tissue rather than to injury as such. In fact the effect was confined to the specific epithelium while the connective tissue stroma remained unaffected.

This demonstration of direct compensatory growth reactions resulting from disturbance of the intracellular-extracellular balance of complementary organ-specific compounds in no way rules out the occurrence of true "functional" hypertrophy as a result of overload; the relative roles played by the two processes will presumably vary from object to object. Moreover, in the endocrine system additional compensatory regulations arise from the reciprocity of hormone relations between different glands.

The direct chemical balance reaction illustrated above may turn out to be a ubiquitous and general principle to which functional and hormonal effects would be merely superimposed. Several considerations indicate its general validity. For instance the observation of an increase in the undamaged liver of one partner of a pair of parabiosed rats following partial removal of the liver of the other partner (Bucher, Scott, and Aub, '51; Wenneker and Sussman, '51) lends itself to the same interpretation.[2] In fact the correlation of liver regeneration rate with blood flow (Flores, '52) and the increase of this rate when the blood is diluted (Glinos and Gey, '52), thus reducing the concentration of our hypothetical liver-"antitemplates" in the circulation, add further support to our interpretation.

[2] Further confirmation has recently been produced by Friedrich-Freksa and Zaki (*Ztschr. Naturf.* 9b, 394-397) who found mitotic spurts in the livers of normal rats injected intraperitoneally with serum from partially hepatectomized animals.

VII. INJECTION OF CELL DEBRIS

Homologous growth effects were also obtained by the direct injection into the embryonic blood stream of triturated cell masses from kidney or liver, either fresh or after freezing and thawing. These experiments, carried out by Gert Andres in our laboratory (in press in *Journal of Experimental Zoology*) and involving a count of 86,000 mitoses, demonstrated that the mitotic ratio of host kidney to host liver is within a day significantly ($P < 0.001$) raised by the injection of kidney material, and lowered by liver material, each organ debris exerting an homologous effect.

Comparable results on a smaller scale have been reported for the orbital glands of rats following intraperitoneal injection of homologous extract (Teir, '52).

VIII. HOMOLOGOUS ORGAN EXTRACT IN TISSUE CULTURE

In the light of our theory, the common observation of a "growth-stimulating" effect of embryo extract on the proliferation of tissue cultures would be accounted for by the fact that, since embryo extract contains cell debris of all organs, the growth of any tissue explanted in it would be favored. In order to put this contention to a test, large series of tissue cultures of kidney and heart were set up in paired media, one containing extract of the complete embryo, the other, extract from which the homologous organ was omitted. The experiments, carried out with Ilse Fischer, showed remarkable differences between the two sets of conditions.

In the kidney experiments (6,300 cultures of 12-day mesonephros and 2,335 cultures of 9- or 17- to 20-day metanephros), the frequency of tubule differentiation was used as a criterion. In the presence of kidney extract this frequency was greatly reduced. Since differentiation in tissue culture is generally conceded to be in some inverse relation to the intensity of proliferation, this result could be interpreted as an homologous growth stimulation by the kidney debris in the medium.

In the heart cultures the differentiation of new myofibrils, evidenced by continued pulsation in successive transfers with subdivision, was used as a sign of depressed growth. In a first series of 978 paired cultures, with and without extract of 5- to 6-day hearts, a large preponderance of pulsation was found in the absence of heart extract, signifying presumably reduced growth. In a later reinvestigation (with Margaret W. Cavanaugh) it was noted, however, that the effect could not be definitely established unless the embryo extract was from embryos older than about 9 days. After this time it made a great deal of difference for heart cultures whether or not heart extract was present in the medium. Such an age effect was previously reported by Gaillard for tissue cultures in general and by Ebert ('51) for the growth stimulation of spleen by chorio-allantoic spleen grafts. Possibly the "growth-promoting" potency ascribed by Hoffman and Doljansky ('39) to heart extract can be related to the fact that their standard assay objects were heart fibroblasts.

An extension of our experiments to tissue cultures of skin and thyroid has thus far given no comparable results, presumably because of the lack of sharp criteria. Even so, the use of tissue culture for the further analysis of these homologous growth interactions between specific cell types and corresponding cell constituents in the culture medium seems to hold much promise. Whether cell compounds in the extract promote homologous growth by being directly incorporated into the corresponding cells, or by neutralizing homologous growth inhibitors in the medium, is still unresolved.

IX. IMMUNOLOGICAL EXPERIMENTS

After these varied reconfirmations of the effect of cell extracts on homologous growth, it became particularly intriguing to resume our original attempts to secure similar effects with antibodies against specific organ extracts. As mentioned above, we had found livers and kidneys of chick embryos, treated with liver- or kidney-antisera, respectively, to be significantly larger than after treatment with non-homologous antisera. A repetition of the liver experiments confirmed the size increase but proved it to be due to hemorrhages from specifically damaged liver vessels. The homology of the action is still evident but its relation to growth is unsubstantiated. Injection of anti-lens serum likewise failed to produce any appreciable effect on the growth of the host lenses (work with Audrey Peterson).

After these failures to detect specific growth alterations from antiorgan sera, it seemed necessary to turn back one step and ascertain whether or not at least the preferential incorporation of anti-organ sera in homologous embryonic organs could be proved. Immumological reactions of specific embryonic tissues to homologous antibodies have been demonstrated (Burke et al., '44; Ebert, '50; Grunwalt, '49), and it was rather obvious to try to demonstrate the selective absorption directly by using isotope-tagged antibodies. Unfortunately five years of continued efforts in that direction have failed to produce the expected results (immunological work with the aid of D. H. Campbell, California Institute of Technology, and Robert Petzold; experiments and isotope assays aided by Gert Andres, Howard Holtzer, Margaret W. Cavanaugh, Evelyn R. Mills, and James Lash).

Antibodies tagged with C_{14}-glycine injected into the yolk sac were concentrated in the embryo, but proved to be of too low titer and radioactivity for a test of selective distribution. We then turned to purified antibody preparations (gamma globulin fraction) tagged with I_{131} according to Pressmann and injected directly into embryonic veins by the technique of Weiss and Andres ('52). Radiation assays proved that the injected material was differentially distributed throughout the embryonic tissues, with blood, heart, and kidneys showing the highest concentrations and other organs following in a certain order. However, there was no evidence of greater absorption of a given organ-antibody by the homologous embryonic organ. It remains undecided whether these failures are due to technical imperfections or actually prove that organ anti-sera are not absorbed in demonstrably larger quantities in the homologous cells of the embryo. It is quite possible, of course, that if growth promotion by these antibodies were of a catalytic nature, even amounts too small to be detectable could exert potent effects.

X. CONCLUSIONS

From this brief survey it seems that the existence of a cell-type specific chemical mechanism of correlating growth processes among homologous cell types must not only be postulated but also also may be regarded as conclusively demonstrated. As for the nature of this mechanism, my own concept or theory, advanced on previous occasions and reiterated here, has proved its value as a guide but must not be taken to have been either proved or disproved. In its current form it is probably too simple to be wholly correct; further work will amend or even replace it. But the general idea of selective chemical communication among cells of identical types by direct exchange of protoplasmic type-specific compounds can hardly be questioned in view of the large evidence in its favor. My earlier detailed suggestion that the complementary systems of growth-catalyzing and growth-repressing compounds are of the antigen-antibody class has found no direct support in our further immunological studies; but it has not been definitely ruled out. Immunoembryology has made vigorous strides of late, but most of that work has been devoted to the detection and tracing by immunological techniques of the appearance of certain antigens during differentiation, e.g. Cooper

('50), Woerdemann ('53), ten Cate ('50), Schechtmann ('52), and Clayton ('53), rather than to the possible instrumental role of these systems as determining and regulatory factors, as I had originally proposed. This latter view, also adopted by Tyler ('47) in an extension of his earlier ideas on "autoantibodies" to problems of growth, therefore remains open to question, but it still deserves to be kept in mind, if only as a model.

BIBLIOGRAPHY

Arataki, M. 1926. *Am. J. Anat. 36,* 437-450.

Brues, A. M., D. R. Drury, and M. C. Brues. 1936. *Arch. Path. 22:* 658-673.

Bucher, N. L. R., J. F. Scott, and J. C. Aub. 1951. *Cancer Res. 11,* 457-465.

Burke, V., N. P. Sullivan, H. Peterson, and R. Weed. 1944. *J. Infect. Dis. 74,* 225-233.

ten Cate, G., and W. J. Van Doorenmaalen. 1950. *Proc. Kon. Ned. Ak. Wetensch. 53,* 894.

Clayton, R. M. 1953. *J. Embryol. Exp. Morph. 1,* 25-42.

Cooper, R. S. 1950. *J. Exp. Zool. 114,* 403-420.

Danchakoff, V. 1916. *Am. J. Anat. 20,* 255-327.

Ebert, J. D. 1950. *J. Exp. Zool. 115,* 351-378.

Ebert, J. D. 1951. *Physiol. Zool. 24,* 20-41.

Flores, N. 1952. *C. R. Soc. Biol. 146,* 589-591.

Glinos, A. D., and G. O. Gey. 1952. *Proc. Soc. Exp. Biol. Med. 80,* 421-425.

Golgi, C. 1882. *Arch. ital. biol. 2.*

Greenman, M. J. 1913. *J. Comp. Neurol. 23,* 479-513.

Grunwaldt, E. 1949. *Texas Reports on Biol. Med. 7,* 270-317.

Guyer, M. F., and E. A. Smith. 1918. *J. Exp. Zool. 26,* 65-82.

Haasler. 1891. *Centralbl. allg. Path. path. Anat. 2,* 809.

Hoffmann, R. S., and L. Doljanski. 1939. *Growth 3,* 61-71.

Kollros, J. J. 1940. *J. Exp. Zool. 85,* 33-52.

Köster, G. 1903. *Neurol. Centralbl. S. 1093.*

Morales, M. F., and F. L. Kreutzer. 1945. *Bull. Math. Biophysics 7,* 15-24.

Nittono, K. 1923. *J. Comp. Neurol. 35,* 133-161.

Pressman, D. 1949. *Cancer 2,* 697-700.

Przibram, H. 1907. *Roux' Arch. 25,* 266-343.

Ribbert, H. 1895. *Roux' Arch. 1,* 69-90.

Ribbert, H. 1904. *Arch. f. Entwicklgmech. 18,* 267-288.

Robertson, O. H. 1954. In publication; personal communication.

Rollason, H. D. 1949. *Anat. Rec. 104,* 263-285.

Schechtman, A. M., and H. Hoffman. 1952. *J. Exp. Zool. 120,* 375-390.

Smadel, J. E. 1936. *J. Exp. Med. 64,* 921-942.

Tamaki, K. 1936. *J. Comp. Neurol. 64,* 437-448.

Teir, H. 1951. *Commentationes Biol. 13,* 1-32.

Teir, H. 1952. *Acta pathol. et microbiol. Scand. 30,* 158-183.

Tyler, A. 1947. *Growth 10* (suppl.), 7-19.

Weiss, P. 1939. *Anat. Rec. 75* (suppl.), 67.
Weiss, P. 1947. *Yale J. Biol. Med. 19*, 235-278.
Weiss, P. 1949. *Chemistry and Physiology of Growth*, pp. 135-186, Princeton Univ. Press.
Weiss, P. 1952. *Publ. Ass. Res. Nerv. Ment. Dis. 30*, 3-23.
Weiss, P., and G. Andres. 1952. *J. Exp. Zool. 121*, 449-488.
Weiss, P., and H. Wang. 1941. *Anat. Rec. 79*, 62.
Wenneker, A. S., and N. Sussman. 1951. *Proc. Soc. Exp. Biol. Med. 76*, 683-686.
Willier, B. H. 1924. *Am. J. Anat. 33*, 67-103.
Woerdeman, M. W. 1953. *Arch. Neerland. Zool. 10* (suppl.), 144-162.
Zeleny, Ch. 1902. *Roux' Arch. 13*, 597-609.
Zeleny, Ch. 1905. *J. Exp. Zool. 2*, 1-102.

XI. SPECIFICITY OF NUCLEAR FUNCTION IN EMBRYONIC DEVELOPMENT

BY ROBERT BRIGGS AND THOMAS J. KING[1]

I. INTRODUCTION

THE STUDY of differentiation may concentrate either on the reactions between cells or groups of cells in the embryo or on events occurring within individual cells as they differentiate. With regard to the intercellular or supracellular reactions, there have been in these Symposia valuable contributions on the problems of induction (Holtfreter, 1951), on growth regulation (Weiss), on the origin of specific substances with antigenic or enzymatic properties (Woerdeman, Schechtman, Shen, Brown), and on other related subjects. In this chapter we will be concerned with intracellular events and in particular with the role of the nucleus in differentiation. Of the several possible approaches to this problem two will be singled out. The first of these involves a study of nuclear differentiation in normal embryos; the second consists of investigations of nucleo-cytoplasmic interactions in amphibian hybrids. As we shall try to show, both of these types of experiment represent approaches to the question of functional specificity in nuclei of embryos.

First, let us take up the question of nuclear differentiation. Stated in its shortest form this is the question of whether or not cell differentiation involves irreversible genetic changes in the nuclei. The problem may of course be stated in other ways. For example we may ask whether the nuclei of, say, muscle cells are irreversibly specialized so that they can participate only in the synthesis of myosin and associated substances. Or is the seat of this specificity of synthesis to be found in some cytoplasmic component, the nuclei remaining genetically unaltered during development and capable of participating in various types of synthesis, depending on the cytoplasmic environment? Other possibilities are that irreversible genetic changes may occur in both nucleus and cytoplasm or in neither.

[1] Institute for Cancer Research and the Lankenau Hospital Research Institute, Philadelphia. The authors' experimental work reported in this chapter has been supported in part by a research grant from the National Cancer Institute of the National Institutes of Health and in part by an institutional grant from the American Cancer Society. The valuable assistance of Miss Marie DiBerardino in the research reported here is gratefully acknowledged.

This problem has concerned embryologists and geneticists for a long time. One of the earliest theories of differentiation was that of Roux and of Weismann, which proposed the nucleus as the principal agent of differentiation. According to this view the nucleus at the beginning of development contains a complete set of determiners of differentiation. During cleavage these are parceled out in a regular way to the various parts of the egg where they later elicit the appropriate types of differentiation. This theory was soon discarded because various experiments showed that the distribution of the nuclei, during the early cleavages at least, could be changed at will without altering the pattern of differentiation. Embryologists have therefore generally adopted the view that the nuclei are equivalent and have regarded the well-known localizations in the egg cytoplasm as the principal agents of differentiation (Wilson, 1925, pp. 1057-1062). The experiments upon which this conclusion was based were indeed beautiful and convincing, but it is worth emphasizing that the actual evidence was obtained only for the earliest phases of development and therefore has definite limitations. For example it says little or nothing about the possible roles of the egg nucleus and follicle cell nuclei in the origin of the cytoplasmic localizations during the growth of the egg. This is a most important problem (see Schultz, 1952) which needs to be mentioned here even though there will be no opportunity to consider it further. A second limitation (and this is the one we shall be concerned with) is that the equivalence of the embryonic nuclei following fertilization has been proved only for the first few cleavages. Whether the nuclei remain genetically identical in the later stages of development or become irreversibly changed as the various parts of the embryo differentiate was not determined.

This problem, which received very little attention for many years, has lately been attacked anew with some of the improved techniques now available in cytology and biochemistry. A new method—that of nuclear transplantation—has also been applied to the problem. In what follows we shall refer briefly to the recent cytological and biochemical work and then will review the results so far obtained by means of nuclear transplantation.

II. CYTOLOGICAL AND BIOCHEMICAL EVIDENCE OF VARIATION AMONG SOMATIC NUCLEI

Let us consider first the cytological work. The main question here is whether clear-cut differences of any kind can be seen among nuclei of different cell types. Even a cursory examination of, say, an advanced

amphibian embryo will reveal definite variations in the morphology of
the interphase nuclei (Fig. 1). For example those of the lens epithelium
are of ordinary appearance while those of the lens proper are enlarged,
oblong, and contain faintly staining diffuse chromatin and a large

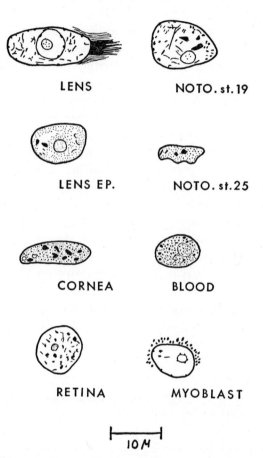

LENS   NOTO. st. 19

LENS EP.   NOTO. st. 25

CORNEA   BLOOD

RETINA   MYOBLAST

10 μ

Fig. 1. Camera lucida drawings showing the differences in the morphology of the interphase nuclei of various tissues of advanced frog embryos (*Rana pipiens*). From Feulgen
and fast green stained sections of 9 mm. embryos (Shumway, stage 23), except for the
notochord nuclei which are from 5 mm. (stage 19) and 11 mm. (stage 25) embryos.

nucleolus of complex structure surrounded by a chromatin-free area.
Several other variations in nuclear structure, as seen in frog embryos,
are illustrated in Fig. 1. Some of these morphological changes may be
irreversible. This would appear to be true of mammalian erythroblast
nuclei which become pycnotic and are lost as the cells mature. However,
in most cells the nuclei do not go through any such radical change, and

[209]

it is impossible at this time to say what the genetic significance of these morphological variations in interphase nuclei may be.

Studies of the chromosomes of various types of cells are of more direct interest in this latter connection. A particularly clear case of chromosomal variation was described long ago by Boveri (1892), who found that the chromosomes of all somatic cells in the Ascaris embryo lose a large part of their chromatin while those of the germ line remain intact. For a long time this was regarded as a relatively rare case—an exception to the rule that cells of various kinds in the same organism contain identical chromosome complements. However, during the past several years numerous other "exceptions" have been discovered, particularly among plants and insects (reviewed by Huskins, 1947; Schultz, 1952). One of the most important of these involves the heterochromatic chromosomal regions, which have important effects on the function of the adjacent genes leading to different expressions in different cells, i.e. to variegation. In some cases, as in Maize, the mitotic mechanism is affected in such a way as to bring about a somatic segregation of the affected genes (McClintock, 1951). In other instances, e.g. Drosophila, the variegation process is not known to include such a mechanism for somatic segregation (Schultz, 1939; 1947). Here the effect of the heterochromatic regions may be primarily on the functional stability of the gene, rendering it susceptible to differences in the cytoplasmic environment in different cells. Gene changes thus induced would act reciprocally upon the cytoplasm, leading to explicit cell differentiation in some such manner as was suggested in general terms by Morgan (1934).

In addition to these effects of heterochromatin on gene function and distribution, there are numerous other instances of variations involving chromosome structure and number. Some of these variations are best seen in the giant chromosomes of insects. For example in Chironomus larvae the chromosomes of the salivary glands, Malpighian tubules, and gut epithelium show differences in degree of polyteny and, more important, the structure of homologous bands show specific differences in the different tissues (Beermann, 1952; Kosswig and Shengün, 1947). In plants also the degree of polyteny or polyploidy may vary as in the cases reported by Huskins and Steinitz (1948) in which differentiation of root tips is accompanied by increases in chromosome number. Similar chromosomal variations occur in certain tissues in mammals, notably in the liver and in many tumors (Hauschka and Levan, 1953; Biesle, Poyner, and Painter, 1942). Finally there have been reports that some somatic cells may undergo reduction divisions (Berger, 1938; Huskins and Steinitz, 1948).

In this connection the recent work of Lindahl (1953) is of special interest to embryologists. According to Lindahl the micromere quartet of the sea urchin embryo undergoes, at the 16 → 32 cell stage, a reduction division producing haploid micromeres in an otherwise diploid embryo.[2] A similar condition exists also in mesenchyme cells of the tail tips of young frog tadpoles (Green, 1953). The majority of these contain the haploid number of chromosomes while other tissues, e.g. liver and epidermis, are diploid, as expected. These results with sea urchin and frog embryos suggest that in some cases a somatic reduction may be involved in differentiation of mesoderm.

In addition to the cytological variations mentioned above there are also biochemical variations among somatic nuclei. Thus nuclei of avian erythrocytes contain hemoglobin, which is lacking in nuclei of other tissues, and the activities of several enzymes are quite different in the nuclei of different tissues (Mirsky, 1951; Stern et al., 1952). Differences are also found in the amounts of histone, protamine, residual protein, and ribonucleic acid present in nuclei of different tissues. The average content of desoxyribosenucleic acid on the other hand is about the same (per chromosome set) in several tissues (Boivin, Vendreley, and Vendreley, 1948; Mirsky and Ris, 1949). However, even here there appear to be significant variations, particularly among nuclei of differentiating embryonic cells (Lison & Pasteels, 1951; Moore, 1952).

This brief survey shows that during recent years an increasing number of cases have been discovered in which there are definite differences among somatic nuclei, and one would guess that there are actually many more differences existing than have so far been detected. Now, what is the significance of these nuclear phenomena for cell differentiation? The biochemical data demonstrate the important fact that there are functional, i.e. enzymatic, differences among somatic nuclei. These differences might depend on differential distributions or activities of genes, but at present there is little or no evidence that they do (Mirsky, 1951). The same may be said about the morphological variations in interphase nuclei. The chromosomal variations on the other hand can be more readily interpreted with regard to their possible roles in differentiation. Thus changes in degree of polyploidy or polyteny may easily result in changes in the relative activity of different genes, and differential segregations

[2] In recent studies Makino and Alfert have failed to find reduction divisions or haploid chromosome numbers in micromeres of the California sea urchin *Strongylocentrotus purpuratus* (personal communication from Dr. Max Alfert). Lindahl's studies were done on the European species *Paracentrotus lividus* and *Strongylocentrotus droebachiensis*.

of parts of chromosomes or differential changes in gene function would provide an obvious mechanism of differentiation (McClintock, 1951; Schultz, 1952). The difficulty in finding out if chromosomal changes of these kinds are part of a general mechanism of differentiation is of course a methodological one. The fact is that in the embryos most commonly studied the number and structure of the chromosomes in various differentiating tissues do not show the striking variations seen in certain insect and plant tissues. Thus the chromosomal variations, if they occur, must be on a very small scale. Refinements in cytological techniques may in time reveal such morphological changes in nuclei and chromosomes as may be associated with differentiation. Meanwhile a new method which promises to give direct evidence concerning nuclear differentiation may now be applied to the problem. This is the method of nuclear transplantation which theoretically should permit us to detect irreversible changes in nuclear function whether they are accompanied by morphologically visible changes or not. A discussion of this method and of the results so far obtained with it is given in the following section.

III. NUCLEAR TRANSPLANTATION STUDIES

A. Background. It has been realized for some time that the most direct way to obtain evidence concerning irreversible changes in nuclei or cytoplasm during differentiation would be to transfer these components from one cell type to another (Ephrussi, 1951; Hämmerling, 1934; Lorch and Danielli, 1950; Schultz, 1943, 1952; Spemann, 1938). Ideally this type of work should proceed in two phases. The first of these is aimed at determining whether there are, in differentiated cells, genetic entities determining cell type and whether these are located in nucleus or cytoplasm or both. Given a demonstration of such entities the next phase would be to determine when and how they arise or come to expression during differentiation.

So far the work on this large problem has been concerned mainly with the first phase and most of it has been done with unicellular organisms, because in these forms nuclear and cytoplasmic elements may be transferred from one cell type to another during natural mating and occasionally also by artificial devices. Theoretically the same type of experiment may be done with differentiated cells of metazoa, each cell type being regarded as a stable clone or variety. Here the problem is more difficult to attack because there is so far as we know no mating among somatic cells. Therefore new techniques must be devised for transferring components from cell to cell in living condition, and work along this line is only beginning.

[212]

Transfers of cytoplasm or cytoplasmic components from cell to cell were accomplished several years ago with unicellular organisms. One of the most elegant experiments of this type was that performed by Hämmerling on the single celled alga *Acetabularia*. This plant is differentiated into a rhizoidal end containing the single nucleus and a long stalk terminating in an umbrella-like structure which shows characteristic differences in different species. Since the over-all length of this giant cell may be 2 to 3 cm. it is possible to do with it several types of experiments on the roles of nucleus and cytoplasm in growth and differentiation (Hämmerling, 1934; Brachet, 1952). The experiment which concerns us most here is one in which Hämmerling transplanted nonnucleated stalk pieces from one species to the nucleated rhizoidal ends of another. The characteristics of the umbrella which then regenerated from the transplanted stalk were determined not by the stalk itself but rather by the nucleated rhizoid to which it had been transplanted. From this result Hämmerling concluded that the form of the umbrella is determined by species-specific nuclear products and that there are in *Acetabularia* no comparable entities in the cytoplasm. Of course numerous other experiments, mostly with single-celled organisms, have since shown that the cytoplasm as well as the nucleus may contain genetic units which play a part in the determination of cell type. The commonly cited examples are the chloroplasts in certain plants, the kinetosomes in ciliates (review by Weisz, 1951), Kappa in Paramecia (Sonneborn, 1947; 1951), and the respiratory mutants involving cytoplasmic units in yeasts (Ephrussi, 1951). Speculations have been made concerning the roles which cytoplasmic units of this type might play in differentiation in Metazoa (Ephrussi, Sonneborn, Weisz, Spiegelman, 1948; Rose, 1952). However, there is as yet no evidence of their existence or activity in embryos.

Let us turn now to the work which has been done on these problems by means of nuclear transplantation. Technically this approach has been difficult because of the fragility of nuclei, which in some cases may be irreversibly damaged if they are merely touched with a glass needle and are always killed if exposed to any medium other than intact cytoplasm. Even so nuclei have been successfully transferred from cell to cell in two different organisms, and it will probably turn out to be feasible to make nuclear transfers in more instances as methods are improved.

The first successful nuclear transfer was accomplished by Comandon and DeFonbrune (1939) using *Amoeba sphaeronucleus*. The operation consisted of placing an enucleated host amoeba in close contact with a nucleated donor and then pushing the nucleus from the donor into the host by means of a blunt microneedle. Thus, the nucleus does not come into contact with the surrounding medium and in this amoeba it is apparently tough enough to withstand the manipulation. Following the transfer the host cell, which was formerly quiescent, is reactivated, indicating that the transplanted nucleus is carrying on at least some of its normal functions.

The work with amoebae was carried considerably further by Lorch and Danielli (1950), who succeeded in showing that following nuclear transfer the host amoebae divide and form mass cultures. Lorch and Danielli also made crosses between two species of amoebae, i.e. transfers of nuclei of *A. discoides* into cytoplasm of *A. proteus* and vice versa. The two species differ in nuclear volume and in the form they assume when migrating. In the successful crosses both of these characteristics were intermediate, indicating an influence of both nucleus and cytoplasm in their determination. However, the apparent absence of sharply defined qualitative differences between the two species makes it difficult to exploit these amoebae in studies of differentiation although they may be put to excellent use in other types of studies of nuclear function (Danielli, 1952; Brachet, 1952; Mazia, 1951).

B. Transplantation of nuclei from embryonic cells to enucleated eggs. In order to get directly at the problem of nuclear differentiation during embryonic development, we set out several years ago to devise a method of transplanting nuclei from cell to cell in the amphibian embryo. The two approaches which were open to us were (1) to attempt to transfer nuclei from one type of differentiated cell to another or (2) to transfer nuclei from differentiated cells back into enucleated unfertilized eggs. The first of these approaches is the same as that used in experiments with unicellular organisms in which the transfers of nuclear or cytoplasmic elements are made between different varieties or species. We decided against trying this type of experiment with embryonic cells because of the technical difficulties. Also it was felt that there were certain theoretical disadvantages involved in testing nuclei by transferring them into cytoplasm which is itself already differentiated. If the cytoplasmic differentiation is in any sense irreversible, then the properties of the transplanted nuclei might not be revealed in such an experiment. Thus for both practical and theoretical reasons it was decided to concentrate on the second type of approach, i.e. the transfer of embryonic cell nuclei back into enucleated eggs.

Before valid tests of this kind could be carried out it was necessary to do two things. First, the developmental potentialities of activated non-nucleated eggs had to be explored. In other words the behavior of the test system (egg cytoplasm without nuclei) had to be known before we could judge the significance of results obtained with experimental combinations (egg cytoplasm with transplanted nuclei). Several previous studies had shown that cleavage and blastula formation may occur in eggs lacking nuclei (Fankhauser, 1934; E. B. Harvey, 1936; Stauffer, 1945). Since these blastulae are always arrested before entering gastrulation, it can be concluded that the formation of the inductive system—the dorsal blastoporal lip and its derivatives—can not occur in the absence of nuclei. However, it was not determined whether the nonnucleated cells also lack the ability to differentiate when exposed to normal inductors. In order to settle this question nonnucleated cells from arrested frog blastulae were grafted to inductive sites on normal embryos. The grafts healed in and survived for periods as long as four days but showed no signs of differentiation (Briggs, Green, and King, 1951). Similar results have also been obtained in recent studies of the development of sea urchin embryos consisting of both nucleated and enucleated blastomeres. The enucleated blastomeres showed limited cleavage but failed completely to differentiate (Lorch, Danielli, and Hörstadius, 1953; Hörstadius, Lorch, and Danielli, 1953). These results show quite definitely that in the absence of nuclei the egg cytoplasm may cleave to a limited extent but is unable to differentiate. Thus it represents an adequate test system for the proposed nuclear transplantation studies. By itself it does nothing, whereas in combination with its normal "undifferentiated" nucleus it can display in the course of development the complete range of differentiations. Therefore if the "undifferentiated" egg nucleus could be replaced by one from a differentiated cell, the nature of the ensuing development should reveal the character of the transplanted nucleus. For example suppose we were to transplant the nucleus from a nerve cell back into an enucleated egg. If this egg then developed into a complete embryo, we could conclude that the transplanted nucleus had not undergone any irreversible change in the course of nerve cell differentiation. On the other hand if the recipient egg was found to be capable of producing only neural structures, then we would have evidence for nuclear differentiation, i.e. for an irreversible change limiting the nucleus to participation in only one type of differentiation.

The second requirement for making valid tests of nuclear differentiation was for a method of transplanting nuclei in living undamaged condition. For various technical reasons the procedure used successfully with amoeba nuclei could not be used with nuclei of embryonic cells. After considerable experimentation a different technique was worked out,

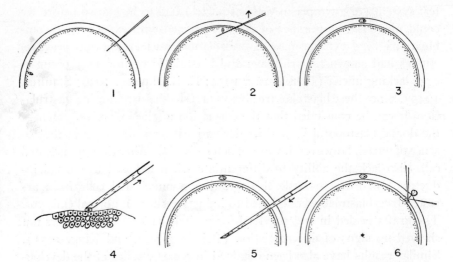

Fig. 2. Diagram showing steps in the procedure for transplanting embryonic nuclei into enucleated eggs of the frog, *Rana pipiens*. Enucleation involves first pricking the egg with a clean glass needle (1). This activates the egg, causing it to rotate so that the nucleus is uppermost. The activation also results in a movement of the cortical pigment granules away from the area over the nucleus, allowing the operator to locate the exact position of the nucleus. The enucleation is performed by placing a glass needle directly beneath the nucleus and then pulling the needle up through the egg surface (2). This produces an exovate which always contains the egg nucleus. The exovate is trapped in the inner jelly coat surrounding the egg (3).

The main steps in the nuclear transfer are shown in 4, 5, and 6. The donor cell is placed on the intact surface coat of a late blastula, which serves as an operating platform. The cell is then carefully drawn up into a micropipette in such a way as to break the cell surface without dispersing its contents (4, 5). The broken cell is injected into an enucleated egg, thus liberating the nucleus into the egg cytoplasm (5, 6). As the micropipette is withdrawn, it pulls the surface coat up through the egg membranes forming a small canal which must be severed with glass needles (6).

In actual dimensions the micropipette is much smaller in proportion to the size of the egg than is indicated in the diagram. Pipette diameters vary from 20 to 50μ, while the diameter of the egg is approximately 1,800μ (1.8 mm).

the main points of which are illustrated in Fig. 2 (Briggs and King, 1952; 1953). Briefly it involves first activating the recipient egg (*R. pipiens*) with a glass needle and then enucleating it by Porter's method (1939). Next the donor cell is carefully pulled up into a micropipette in such a way that the cell surface is broken without dispersing the cytoplasm. The broken cell is then injected into the enucleated egg. When this is done properly the nucleus is protected by its own cytoplasm until the moment when it is liberated into the cytoplasm of the recipient egg. This protection is essential because exposure of the nucleus to any of the artificial media so far tried promptly kills it (Briggs and King, 1953, and unpublished). The volume of the cytoplasm which must be injected

PLATE I. Photograph of apparatus used in carrying out nuclear transplantations. The micropipette (see Fig. 2) is carried on an Emerson micromanipulator. The pressure within the pipette is controlled by the syringe mounted on the left side of the microscope. The chain on the left focusing knob of the microscope passes through a slit in the table top to a foot-focusing device below. The device (not shown in photograph) consists of a right angle gear, one side of which carries a small sprocket for the lower end of the chain and the other side of which is attached to a rod leading to a horizontal wheel mounted close to the floor. The microscope is focused by moving the wheel with either foot. (This focusing device is a copy of one that has been used for some years in the Biology Department at Swarthmore College). The transformer and variac mounted on the shelf behind the microscope control the two lights which are required to provide adequate illumination of the operating field.

along with the nucleus is very small (ca. 1/100,000) compared with the volume of the recipient egg. Even so the necessity of including it introduces into the method a complication which can not be avoided since it will probably never be possible to guarantee the complete removal of cytoplasm from nuclei in such experiments. The complication is not a serious one, however, because controls for the effect of the cytoplasm can be made where necessary.

Using the method outlined above we have so far carried out tests for nuclear differentiation in blastulae and in early and late gastrulae. Nuclei of animal hemisphere cells of blastulae are relatively easy to transplant. About one-third of the recipient eggs cleave normally and form complete blastulae, the majority of which gastrulate and develop into complete embryos. Nuclei from the corresponding part of early gastrulae are somewhat more difficult to transplant but give essentially the same result. Again the successful transfers lead to normal cleavage of the recipient eggs, followed by gastrulation and the development of complete embryos (Briggs and King, 1953). It is clear from these results that the nuclei of animal hemisphere cells of late blastulae and early gastrulae are not differentiated—when transferred back into enucleated eggs they can still participate in all types of differentiation. This result is more or less expected since the cells from which the nuclei are taken are still totipotent, i.e. capable of any kind of differentiation.

The next phase of the investigation involved transferring nuclei from the late gastrula (stage 12, Shumway) to enucleated eggs. At this stage the dorsal lip of the blastopore has given rise to the chordamesoderm, which in turn has induced the first phase of neural differentiation in the overlying ectoderm. As donor cells we chose those of the presumptive chorda and presumptive neural plate. Nuclei from these cells were transferred to enucleated eggs in the usual way. These transfers turned out to be difficult to do and gave rise to normal cleavage in a relatively small proportion of the recipient eggs (9 per cent) compared with 21 per cent for nuclei of early gastrulae and 34 per cent for those of late blastulae. In other words the proportion of nuclear transfers that turn out to be successful falls off steadily as the age of the donor increases. There is evidence that this decrease in transplantability is not due to intrinsic changes in the nuclei but rather is to be attributed to an increased likelihood of damage to the nuclei in the course of the transplantation. Between the blastula and gastrula stages there is considerable reduction in cell size. This means that the amount of cytoplasm protecting the nucleus during the transplantation is reduced and that there is a much greater chance of the nucleus being damaged by the medium. On the basis of this and other considerations it appears that the majority if not all of the failures are probably due to damage accidentally inflicted on the nucleus in the course of the transplantation.

Despite the difficulties mentioned above a significant proportion of the attempted transfers of presumptive chorda and medullary plate nuclei were successful, giving rise to normal cleavage in the recipient eggs. The normally cleaved eggs developed into complete blastulae. Some of these were arrested as blastulae but the majority gastrulated and continued on to neurula or postneurula stages. There were no significant differences between the eggs developing with nuclei derived from presumptive chorda and those developing with nuclei of presumptive neural plate. Both groups of embryos displayed differentiation of the central nervous system and its derivatives and of mesodermal and endodermal structures (King and Briggs, unpublished).

These results show that the nuclei of at least some of the cells in question are definitely not differentiated. Now the pertinent question is whether the cells themselves are differentiated. On the basis of the usual embryological tests of transplantation and explantation, we are fairly sure that the chordamesoderm and the overlying presumptive neural plate are determined to differentiate into mesodermal and neural organs respectively. However, it is not at all certain whether this determination is a property of the individual cells making up these structures or is to be regarded rather as a property of the whole mass, the individual cells still being at this stage undifferentiated. Recent results of Grobstein and Zwilling (1953) indicate that the latter of these two possibilities is the correct one. These investigators have shown that the differentiation of cultured explants of pieces of chick blastoderm (definitive primitive streak stage) depends on the degree to which the cells are made to disperse. For example a piece of blastoderm of a given size will differentiate into neural tissue when explanted whole but will fail to differentiate or will differentiate poorly if it is divided into eighths or sixteenths prior to culture and subsequent grafting to the chorioallantois. In both the chick blastoderm and the mouse embryonic shield previously studied by Grobstein (1952) it appears that organ determination occurs while the constituent cells are still undifferentiated or in a labile stage of differentiation. The same may very well be true in the late gastrula of amphibians, determination of the chordamesoderm and medullary plate being at first a property of regions and not of cells. Our results indicate that this initial phase of differentiation does not require specific irreversible changes in nuclear function nor does it involve the elaboration of cytoplasmic genetic units capable of directing differentiation when transferred back into uncleaved eggs. Apparently the manifestations of differentiation in the gastrula still depend specifically on the localization of cytoplasmic materials present in the egg at the beginning of development. Therefore the role of the nucleus in this phase of differentiation can probably be studied only during the growth of the oocyte when the cytoplasmic materials are being synthesized under the influence of egg and follicle cell nuclei. Exactly how long this initial phase of differentiation lasts is not yet clear. On the basis of Grobstein's results and of Holtfreter's (1947) studies of the differentiation of single cells isolated from neurulae, it is to be expected that at some point during neurulation differentiation may become fixed in individual cells. The important task confronting people who concern themselves with this problem is now to determine when the individual cells actually do become differentiated and whether this differentiation depends on irreversible genetic changes in the nucleus or elsewhere in the cell.

IV. NUCLEOCYTOPLASMIC INTERACTIONS IN LETHAL
    HYBRIDS (AMPHIBIA)

In the past, one of the principal attacks on the problem of nucleocytoplasmic interaction in development has been made through studies of hybrids (see reviews by P. Hertwig, 1936; Hörstadius, 1936; Baltzer, 1952; Fankhauser, 1952). It appeared to us that in this field of investigation it should be possible to do certain types of experiments by means of nuclear transplantation that could not be done in any other way. Two such investigations have recently been carried out. The first of these has to do with the development of hybrids between two species of frogs, *Rana pipiens* ♀ × *Rana catesbeiana* ♂ (Rugh and Exner, 1940; Moore, 1941; Briggs, Green, and King, 1951). This hybrid along with others of the same type is of particular embryological interest because it develops normally to the beginning of gastrulation, is then abruptly arrested, and later cytolyses without developing further. In other words the incompatibility between nucleus and cytoplasm is expressed just at the beginning of the main morphogenetic events of development and in this particular cross results in an irreversible block to differentiation. The question concerning this hybrid which we set out to answer was the simple one of whether the irreversible block to differentiation involves irreversible changes in the nuclei. Tests for such changes were carried out in the following way: first, the hybrids were

[219]

produced by inseminating *pipiens* eggs with *catesbeiana* sperm. In some cases androgenetic haploid hybrids were obtained by removing the egg nucleus following insemination with the foreign sperm. Nuclei were then transferred from the hybrids back into enucleated *pipiens* eggs at the time of onset of developmental arrest (26 hours) and at intervals preceding and following that time. If at one or another of these times the hybrid nuclei have undergone an irreversible change, this change should be revealed by an alteration in the development of the recipient eggs. Actually the results of these tests showed that there were no irreversible nuclear changes associated with the arrest of development at the beginning of gastrulation (King and Briggs, 1953). Nuclei of arrested hybrid gastrulae were capable, when transferred back into enucleated *pipiens* eggs, of participating again in the whole course of hybrid development (see Fig. 3). However, beginning at some 10 to 20 hours following the onset of arrest, a radical change occurred in the nuclei which rendered them incapable of eliciting normal cleavage in the recipient eggs. This loss of transplantability coincided with the first appearance of cytological abnormalities, including such phenomena as bridged anaphases, clumped metaphases, unequal chromosome distributions, and vacuolated interphases. These abnormalities are similar to some of those described by Baltzer (1952) in lethal newt hybrids.

The main outcome of the experiment described above is that the nuclei of the hybrids are unchanged at the time when development is arrested and for several hours thereafter. The incompatibility resulting in this arrest is therefore not expressed as a cumulative effect of the foreign cytoplasm on the nucleus nor as a sudden irreversible nuclear change at the beginning of gastrulation. Rather it is to be regarded as a failure of the nucleus, in combination with the foreign cytoplasm, to acquire the specific properties necessary for gastrulation to proceed.

These results pose two problems of general interest. The first is, what are the specific properties, synthetic systems, etc. which begin to operate in normal embryos during gastrulation but fail to appear in the hybrids? In the cross we have studied, this deficiency appears to be irreversible in the sense that it can not be corrected by grafting the hybrid cells to normal embryos. However, in certain other crosses which are also arrested at the beginning of gastrulation the developmental block can be overcome by grafting (Lüthi, 1938; Brachet, 1944; Moore, 1947, 1948). At first sight this difference in the results of the grafting experiments suggests that there may be an important biochemical difference between the two types of hybrids, the cells of the irreversibly blocked embryos being un-

[220]

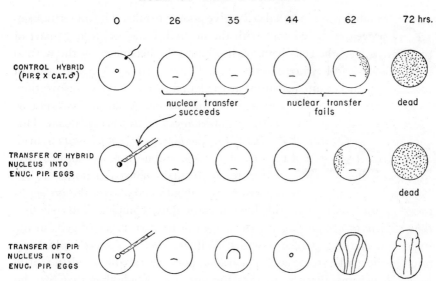

Fig. 3. Diagram illustrating the results of transplanting nuclei from arrested hybrids (*R. pipiens* ♀ × *R. catesbeiana* ♂) into enucleated eggs of *R. pipiens*.

The top row of figures shows the development of normally produced hybrids which develop normally to the beginning of gastrulation and are then arrested. Transfer of nuclei from the arrested hybrids at first leads to typical hybrid development of the recipient eggs (second row of figures). Later on, at about 44 hours, the nuclear transfers fail to elicit cleavage in the recipient eggs. The same result, with minor variations, was obtained when nuclei of androgenetic haploid hybrids [*pipiens* ♀ × *catesbeiana* ♂] were transferred back into enucleated *pipiens* eggs.

The bottom row of figures shows by way of comparison the result of transferring *pipiens* nuclei (late blastula or early gastrula) into enucleated *pipiens* eggs.

able to synthesize some nondiffusible component which is essential for gastrulation, while the deficiency in the reversibly arrested hybrids is in the form of a diffusible substance which can be obtained from the normal host (see Brachet, 1952). While this interpretation is not disproved it now appears less well founded than it did at first, principally as a result of recent cytological observations, indicating that at least some of the reversibly blocked hybrids are what Baltzer (1952) refers to as partial lethals. After the onset of developmental arrest these hybrids consist of a mixture of cells which display nuclear abnormalities and cells with normal looking nuclei. In the ordinary course of events the "lethal" cells are eliminated into the blastocoel while the normal ones persist for some time even though the embryo as a whole fails to develop, possibly because of the presence of the dead cells. Late grafts made after the elimination of the lethal cells survive and differentiate on the normal hosts. Early grafts, made before the appearance of the cytologi-

[221]

cal abnormalities in the lethal cells, give poorer results. The abnormalities are not prevented by contact with the normal tissue and a large part of the graft is lost, the cells being cast off into the blastocoel of the normal host (Baltzer and Schönmann, 1951; Baltzer, 1952). On the basis of this analysis it is quite certain that the lethal effect of the hybridization, as expressed in the cells which show nuclear abnormalities, can not be prevented or reversed by diffusible substances from normal hosts. This conclusion is confirmed by our study of the *pipiens* × *catesbeiana* hybrid. Here virtually all of the cells display nuclear abnormalities and the hybrid tissue dies whether it is grafted to normal embryos or not.

As a result of the cytological findings mentioned above, the problems posed by the "lethal" hybrids have become more complex. There is first the problem of accounting for the origin of the two types of cells in the hybrids displaying partial lethality. With respect to the cells which retain the normal cytological appearance and are capable of differentiation on grafting, there is further the question of whether they are deficient in certain diffusible substances which they receive from the host or are actually not different from normal cells in their requirements. Finally with respect to the "lethal" cells there is the problem of determining the nature of their deficiencies in relation to their failure to survive and differentiate. In this connection we may mention the cytochemical studies of Brachet (review, 1952), who has shown that the distribution of nucleic acids may be quite abnormal in the hybrid cells. The RNA of the cytoplasm is not increased in amounts as it is in normal gastrulae while the nuclei appear to contain abnormally large amounts of RNA. Chemical measurements by Steinert (1953) show that the total synthesis of RNA in the arrested hybrids is abnormally low and that there is a correlated reduction in the utilization of presumed RNA precursors —the free bases, guanine and hypoxanthine. Similar observations have been made with respect to oxygen consumption (Barth, 1946), glycogen utilization (Gregg, 1948), and rate of phosphorylation (of ATP) (Barth and Jaeger, 1947). These activities also remain at levels characteristic of the blastula instead of showing the increases which normally occur during gastrulation. The substances and activities mentioned above are thus among the ones which the hybrid fails to display and which are, in ways as yet unspecified, associated with synthetic processes occurring during development beyond the gastrula stage (see review by Brachet, 1952).

The second problem of general interest stems from the fact that *prior* to gastrulation the development of the lethal hybrids is quite normal.

This means, for example, that a single haploid nucleus in the foreign cytoplasm is duplicated many times, producing thousands of nuclei within a day or so. And this process could probably go on indefinitely if we were to keep on transplanting nuclei from hybrid blastulae back into enucleated eggs. The explanation of this behavior may be that the precursors necessary for such rapid synthesis of chromosomal nucleo-protein during cleavage exist ready-made in the egg cytoplasm at the beginning of development (see Zeuthen, 1951). It is further necessary to think that these precursors are unspecific in order to account for the fact that the foreign chromosomes, in our case the *catesbeiana* chromosomes, are produced as rapidly as are the homologous *pipiens* chromosomes.

Of late it appears that these precursors may be something more than hypothetical. For example Hoff-Jørgensen and Zeuthen (1952) have given evidence of cytoplasmic desoxyribosides in the frog's egg, which they suggest may represent precursors for the formation of nuclear nucleoprotein during cleavage. More recently Levenbook (unpublished) has found significant amounts of a variety of possible nucleic acid precursors in the unfertilized Drosophila egg. Work on these substances is only beginning so it is impossible to say how they may be involved in the rapid formation of new sets of chromosomes during cleavage. However, their existence is very suggestive.

In order to see how unspecific this capacity for the utilization of chromatin precursors might be, we have recently made one rather far-fetched cross between a frog and a newt by transplanting nuclei from blastulae of the Japanese newt *Triturus pyrrhogaster* into enucleated eggs of *Rana pipiens*. Fifty-four such transfers were made and almost half (26) of the recipient eggs cleaved. The cleavage followed an approximately normal pattern at first except for the fact that the furrows usually did not extend completely around the vegetal hemisphere. Later on the cleavage in most eggs was restricted to a part of the animal hemisphere, the pattern resembling that previously found in our study of the development of activated "achromosomal" frogs' eggs (Briggs, Green, and King, 1951). Since the partial blastulae so produced were quite certainly not capable of further development, they were fixed and sectioned. A study of the sections showed that portions of the blastulae were genuinely cleaved into cells having the size and form of those of a normal mid- to late blastula. The majority of the cells did not contain chromatin. However, chromatin derived from the injected nucleus was found in both the remaining cells and in the uncleaved vegetal

[223]

part of the egg. Most important, this chromatin was greatly increased in amount relative to the amount found in the single nucleus that was injected. It was organized into masses or "nuclei" which usually did not show chromosomal structure although in a few cases there was evidence that the newt chromosomes persisted for awhile in the frog cytoplasm.

From these results we can make the following conclusions. In the first place the newt nucleus provides a cleavage center or centrosome which functions relatively normally in frog cytoplasm, inducing it to go through a series of regular cleavages. Secondly, although the newt chromatin is not normally organized or distributed during cleavage in the frog egg cytoplasm, it is increased greatly in amount and is in an abnormal fashion fairly widely distributed in the frog cytoplasm during the first several hours of development. This indicates that the nucleoprotein precursors present in the frog egg are unspecific in that they can be utilized in the synthesis of chromatin of a quite foreign species.

V. SUMMARY

In summary we should like to emphasize the following points. The first of these is the one just described, namely that nuclei and cytoplasm of different species or even of different orders cooperate in cleavage and blastula formation. It has been suggested that this may be due to the presence of nucleoprotein and other precursors in the egg cytoplasm, which may be utilized by the foreign nuclei in the formation of cleavage centers and of new sets of chromosomes or masses of chromatin. Some possible precursors of this sort are already known. Further study of their origin during the growth of the oocyte, and of their fate during cleavage, should contribute to an understanding of the biochemistry of early development and may well provide information, not so readily obtainable in other types of material, on the general problem of nucleoprotein synthesis.

The second main point has to do with the sudden development of specificity of nucleocytoplasmic interaction at the beginning of gastrulation. In general only closely related species give hybrids that go through this and later phases of development. In crosses like that between *Rana catesbeiana* and *R. pipins* development stops suddenly in late blastula or early gastrula stages. We have shown in our nuclear transplantation experiments that this arrest of development is not due to nuclear damage but rather is the result of a failure of the nuclei in the foreign cytoplasm to acquire certain hypothetical properties necessary for development to proceed. A survey of the literature indicates that certain synthetic systems, which are normally elaborated or activated during gastrulation, are somehow blocked in the lethal hybrids, the chemistry of the block being as yet unknown.

These points refer to specificity of function of nuclei and cytoplasm of different species. There is finally the question of the specialization or differentiation of nuclei within a given organism. Such differentiation apparently does not occur by late gastrula stage, according to our nuclear transplantation studies. It appears from the literature that the individual cells may not be differentiated at this stage and that the determination of the chordamesoderm and medullary plate as areas still depends primarily on the localization of cytoplasmic materials present in the egg at the beginning of development. In postgastrula stages it is known that nuclei in different tissues may assume different forms and may also differ from each other biochemically. However, it remains to be determined whether these variations are mere modulations or represent instead genuine nuclear differentiations in the form of irreversible genic changes.

BIBLIOGRAPHY

Baltzer, F. 1952. The behaviour of nuclei and cytoplasm in amphibian inter-specific crosses. *Sym. Soc. Exp. Biol. 6,* 230-242.

Baltzer, F., and W. Schönmann. 1951. Ueber die Letalität des Bastards Triton palmatus ♀ × Salamandra atra ♂. *Rev. suisse de Zool. 58,* 495-502.

Barth, L. G. 1946. Studies on the metabolism of development. *J. Exp. Zool. 103,* 463-486.

Barth, L. G., and L. Jaeger. 1947. Phosphorylation in the frog's egg. *Physiol. Zool. 20,* 133-146.

Beermann, W. 1952. Chromomerenkonstanz und spezifische Modifikationen der Chromosomenstruktur in der Entwicklung und Organdifferenzierung von Chironomus tentans. *Chromosoma 5,* 139-198.

Berger, C. A. 1938. Multiplication and reduction of somatic chromosome groups as a regular developmental process in the mosquito, *Culex pipiens. Carnegie Inst. Washington Pub. 496,* 209-232.

Biesele, J. J., H. Poyner, and T. S. Painter. 1942. Nuclear phenomena in mouse cancers. *Univ. of Texas Pub. 4243.*

Boivin, A., R. Vendreley, and C. Vendreley. 1948. L'acide désoxyribonucléique du noyau cellulaire, dépositaire des caractères héréditaires; arguments d'ordre analytique. *Compt. Rend. Acad. Sci. 226,* 1061-1063.

Boveri, Th. 1892. Die Entstehung des Gegensatzes zwischen den Geschlechtszellen und den somatischen Zellen bei Ascaris. *Sitzber. Ges. Morph. Physiol. (München) 8.*

Brachet, J. 1944. Acides nucléiques et morphogénèse au cours de la parthénogénèse, la polyspermie et l'hybridation chez les anoures. *Ann. Soc. Roy. Zool. Belg. 75,* 49-74

Brachet, J. 1952. The role of the nucleus and the cytoplasm in synthesis and morphogenesis. *Symp. Soc. Exp. Biol. 6,* 173-200.

Briggs, R., E. U. Green, and T. J. King. 1951. An investigation of the capacity for cleavage and differentiation in *Rana pipiens* eggs lacking "functional" chromosomes. *J. Exp. Zool. 116,* 455-500.

Briggs, R., and T. J. King. 1952. Transplantation of living nuclei from blastula cells into enucleated frogs' eggs. *Proc. Nat. Acad. Sci. 38,* 455-463.

Briggs, R., and T. J. King. 1953. Factors affecting the transplantability of nuclei of frog embryonic cells. *J. Exp. Zool. 122,* 485-506.

Comandon, J., and P. deFonbrune. 1939. Greffe nucléaire totale, simple ou multiple, chez une Amibe. *Compt. Rend. Soc. Biol. 130,* 744-748.

Danielli, J. F. 1952. Separate nuclear and cytoplasmic actions of drugs. *Nature 170,* 1042-1044.

Ephrussi, B. 1951. Remarks on cell heredity. *Genetics in the 20th Century,* pp. 241-262. Macmillan Co., New York.

Fankhauser, G. 1934. Cytological studies on egg fragments of the salamander Triton. IV. The cleavage of egg fragments without the egg nucleus. *J. Exp. Zool. 67,* 349-394.

Fankhauser, G. 1952. Nucleo-cytoplasmic relations in amphibian development. *Int. Rev. Cytology 1,* 165-194. Academic Press, New York.

Green, E. U. 1954. On the regular occurrence of the haploid number of chromosomes in mesenchymal cells of the tail tip of *Rana pipiens* tadpoles. *Nature 172,* 766.

Gregg, J. R. 1948. Carbohydrate metabolism of normal and of hybrid amphibian embryos. *J. Exp. Zool. 109,* 119-133.

Grobstein, C. 1952. Effect of fragmentation of mouse embryonic shields on their differentiative behavior after culturing. *J. Exp. Zool. 120,* 437-456.

Grobstein, C., and E. Zwilling. 1953. Modification of growth and differentiation of chorioallantoic grafts of chick blastoderm pieces after cultivation at a glass-clot interface. *J. Exp. Zool. 122,* 259-284.

Hämmerling, J. 1934. Über genomwirkungen und Formbildungsfähigkeit bei Acetabularia. *Arch. Entw.-mech. Org. 132,* 424-462.

Harvey, E. B. 1936. Parthenogenetic merogony or cleavage without nuclei in *Arbacia punctulata. Biol. Bull. 71,* 101-121.

Hauschka, T. S., and A. Levan. 1953. Inverse relationship between chromosome ploidy and host-specificity of sixteen transplantable tumors. *Exp. Cell Res. 4,* 457-467.

Hertwig, P. 1936. Artbastarde bei Tieren. *Handbuch der Vererbungswissenschaft Bd. 2,* 1-140.

Hoff-Jørgensen, E., and E. Zeuthen. 1952. Evidence of cytoplasmic desoxyribosides in the frog's egg. *Nature 169,* 245.

Holtfreter, J. 1947. Changes of structure and the kinetics of differentiating embryonic cells. *J. Morph. 80,* 57-92.

Holtfreter, J. 1951. Some aspects of embryonic induction. *Growth Suppl. 10,* 117-152.

Hörstadius, S. 1936. Studien über heterosperme Seeigelmerogone nebst Bemerkungen über einige Keimblattchimären. *Mém. Musée R. Hist. Natur. Belg.*, Ser. 2, Fasc. 3, pp. 803-880.

Hörstadius, S., I. J. Lorch, and J. F. Danielli. 1953. The effect of enucleation on the development of sea urchin eggs. II. Enucleation of animal or vegetal halves. *Exp. Cell Res. 4*, 263-274.

Huskins, C. L. 1947. The subdivision of the chromosomes and their multiplication in non-dividing tissues : possible interpretations in terms of gene structure and gene action. *Am. Nat. 81*, 401-434.

Huskins, C. L., and L. M. Steinitz. 1948. The nucleus in differentiation and development. I. Heterochromatic bodies in energic nuclei of Rhoeo roots. *J. Hered. 39*, 35-44.

Huskins, C. L., and L. M. Steinitz. 1948. The nucleus in differentiation and development. II. Induced mitoses in differentiated tissues of Rhoeo roots. *J. Hered. 39*, 67-77.

King, T. J., and R. Briggs. 1953. The transplantability of nuclei of arrested hybrid blastulae (*R. pipiens* ♀ × *R. catesbeiana* ♂). *J. Exp. Zool. 123*, 61-78.

Kosswig, C., and A. Shengün. 1947. Intra-individual variability of chromosome IV of Chironomus. *J. Hered. 38*, 235-239.

Lindahl, P. E. 1953. Somatic reduction division in the development of the sea urchin. *Nature 171*, 437-438.

Lison, L., and J. Pasteels. 1951. Études histophotométriques sur la teneur en acide désoxyribonucléique des noyaux au cours du développement embryonnaire chez l'oursin *Paracentrotus lividus*. *Arch. Biol. 62*, 1-43.

Lorch, I. J., and J. F. Danielli. 1950. Transplantation of nuclei from cell to cell. *Nature 166*, 329.

Lorch, I. J., J. F. Danielli, and S. Hörstadius. 1953. The effect of enucleation on the development of sea urchin eggs. I. Enucleation of one cell at the 2, 4, or 8 cell stage. *Exp. Cell Res. 4*, 253-262.

Lüthi, H. R. 1938. Die Differenzierungsleistungen von Transplantaten der letalen Bastardkombination Triton ♀ × Salamandra ♂ . *Arch. Entw.-mech. Org. 138*, 423-450.

Mazia, D. 1952. Physiology of the cell nucleus. *Modern Trends in Physiology and Biochemistry*, pp. 77-122. The Academic Press, New York.

McClintock, B. 1951. Chromosome organization and genic expression. *Cold Spring Harbor Symp. Quant. Biol. 16*, 13-47.

Mirsky, A. E. 1951. Some chemical aspects of the cell nucleus. *Genetics in the 20th Century*, pp. 127-153. Macmillan Co., New York.

Mirsky, A. E., and H. Ris. 1949. Variable and constant components of chromosomes. *Nature 163*, 666-667.

Moore, B. C. 1952. Desoxyribose nucleic acid in embryonic diploid and haploid tissues. *Chromosoma 4*, 563-576.

Moore, J. A. 1941. Developmental rate of hybrid frogs. *J. Exp. Zool. 86*, 405-422.

Moore, J. A. 1947. Studies in the development of frog hybrids. II. Competence of the gastrula ectoderm of *Rana pipiens* ♀ × *Rana sylvatica* ♂ hybrids. *J. Exp. Zool. 105*, 349-370.

[227]

Moore, J. A. 1948. III. Inductive ability of the dorsal lip region of *Rana pipiens* ♀ X *Rana sylvatica* ♂ hybrids. *J. Exp. Zool. 108,* 127-154.

Morgan, T. H. 1934. *Embryology and Genetics.* Columbia University Press, New York.

Porter, K. R. 1939. Androgenetic development of the egg of *Rana pipiens. Biol. Bull. 77,* 233-257.

Rose, S. M. 1952. A hierarchy of self-limiting reactions as the basis of cellular differentiation and growth control. *Am. Nat. 86,* 337-354.

Rugh, R., and F. Exner. 1940. Developmental effects resulting from exposure to x-rays. II. Development of leopard frog eggs activated by bullfrog sperm. *Proc. Am. Phil. Soc. 83,* 607-619.

Schultz, J. 1939. The function of heterochromatin. *Proc. 7th Int. Gen. Cong.,* 257-262.

Schultz, J. 1943. Personal communication.

Schultz, J. 1947. The nature of heterochromatin. *Cold Spring Harbor Symp. Quant. Biol. 12,* 179-191.

Schultz, J. 1952. Interrelations between nucleus and cytoplasm: problems at the biological level. *Exp. Cell Res. Suppl. 2,* 17-43.

Sonneborn, T. M. 1947. Developmental mechanisms in Paramecium. *Growth Suppl. 11,* 291-307.

Sonneborn, T. M. 1951. The role of the genes in cytoplasmic inheritance. *Genetics in the 20th Century,* pp. 291-314. Macmillan Company, New York.

Spemann, H. 1938. *Embryonic Development and Induction.* Yale University Press.

Spiegelman, S. 1948. Differentiation as the controlled production of unique enzymatic patterns. *Soc. Exp. Biol. Symp. 2,* 286-325. Academic Press, New York.

Stauffer, E. 1945. Versuche zur experimentellen Herstellung haploider Axolotl-Merogone. *Rev. suisse de Zool. 52,* 231-327.

Steinert, M. 1953. Métabolisme de l'acide ribonucléique dans l'oeuf d'amphibien traité au dinitrophenol. *Biochimica et Biophysica Acta 10,* 427-431.

Stern, H., V. Allfrey, A. E. Mirsky, and H. Saetren. 1952. Some enzymes of isolated nuclei. *J. Gen. Physiol. 35,* 559-578.

Weisz, P. B. 1951. A general mechanism of differentiation based on morphogenetic studies in ciliates. *Am. Nat. 85,* 293-311.

Wilson, E. B. 1925. *The Cell in Development and Heredity* (3rd ed.). Macmillan Co., New York.

Zeuthen, E. 1951. Segmentation, nuclear growth, and cytoplasmic storage in eggs of Echinoderms and Amphibia. *Pub. Staz. Zool. Napoli Suppl. 23,* 47-69.

INDEX

Ablastin, inhibits parasite reproduction, 165-6

Acetabularia, nuclear transplantation in, 213

Achlya, sex hormones in, 121-5; fertilization in, 123

Acid phosphatase, in plant growth, 97-105, 111

Acquired immunity, 158; produced by antibodies, 159; superimposed upon innate immunity, 163; produced by ablastin, 165-6; relation of successive parasite infections to, 166-7

Actin, 44

Adaptive enzymes, in induction and differentiation, 46-7

Adenosinetriphosphatase, and myosin, 75, 78; in developing chick and rat muscle, 78; development of, 78-9; relation to microsomes, 79; in differentiation and morphogenesis, 79-80

Agglutination, mechanism of, 152-3; spatial proximity as a factor in, 153; *see also* Serologic agglutination

Amoeba, nuclear transplantation in, 214

Anterior chamber transplants, 178-80; of adult tissues, 178; of embryonic tissues, 179; homologous transplants, 179; differentiation and organization in, 179; association of embryonic tissues in, 180; heterologous transplants, 182-3; of cancer tissue, 185-6; the problem of compatibility and noncompatibility in, 190-1; *see also* Transplantation

Antibodies, synthesis of, 14; immobilization of microorganisms by, 141-3; cause exudation by microorganisms, 144-7; affect growth pattern, 149-51; and death of parasites, 163; in malarial infection, 163-4; in successive parasite reinfections, 166-7; as parasiticidal substances, 168; effects on development of embryonic organs, 197-8; against specific organ extracts, 203-4; tagged with isotopes in a study of growth, 203-4; *see also* Antiserum

Antibody-forming mechanism, deficiency of in embryos, 15

Antigenic factors, in blood of chickens, 62; in blood of cattle, 62-4; *see also* Antigenic specificities

Antigenic mutants, separation of in bacteria and protozoa, 144

Antigenic specificities, due to gene interaction, 57-60; in species hybrids, 58-9; in blood cells of cattle, 62-6; in heterozygotes of birds, 60-1; *see also* Multiple antigenic specificities of genes

Antigens, epigenicity of in embryonic development, 6; uptake by embryo, 15; in chick development, 39-41; in amphibian development, 41; of lens and iris, 44; relation to induction, 45; *see also* Cellular antigens

Anti-organ serum, to lens, 42-3, 203; problem of preferential incorporation in homologous embryonic organs, 203

Antiserum, immobilizes microorganisms, 141; causes exudation by microorganisms, 144; causes loss in volume of paramecia and tetrahymena, 145; affects growth pattern, 149-51; *see also* Antibodies

Antitemplates, 198; with respect to self-limiting growth, 199; relation to templates, 199-200

Arginine, in plant roots, 104

Autosynthetic serum proteins, 13

Biochemical mosaic, concerned with cellular differentiation, 195-6

Blastema, in limb regeneration, 44; reactions to antisera, 44-5

Blastocyst, permeability of, 18-19

Blood-clotting mechanisms, in embryo and adult, 8-9

Blood plasma proteins, epigenesis of, 8-13

Blood types in cattle and reciprocal transplantation of fetal tissue, 67-8

Boeck's sarcoid, 181

Brain tissue, transplantation of, 188-9

Brain tumors, heterotransplantability of, 187

Breast papilloma, 183, 185

Briggs, 207

Brown, 93

Brown-Pearce rabbit cancer, 184, 191

Cancer tissue, transplantation of, 185; development of autonomy after heterologous transplantation, 185-6

Cattle twins, reciprocal transplantation of fetal tissue in, 67-8; blood types and reciprocal transplantation in, 68

Cell debris, injection of and homologous growth effects, 202

Cell surface, undergoes molecular changes during induction and differentiation, 48

Cellular antigens, and dominant genes, 56; in chick embryos, 60; in embryonic development, 66-7; in cattle, 67; *see also* Antigens

Chlamydomonas, sexuality in, 128

Cholinesterase, in the central nervous system, 84; in neural differentiation, 85-6; in contrast with cytochrome and succinic oxidase, 86-7; in differentiation of the optic system, 89-90

Chorio-allantoic grafts, of mammalian tissue, 24; in a study of factors controlling growth, 197, 203

Chromosomes, variation among different cells, 210-11; giant, 210; polyploidy in, 210-11; biochemical variations of, 211; in lethal hybrids of amphibia, 219-23; nucleic acid precursors of, 223

Cilia, changes during immobilization reaction in protozoa, 142-3

Cleveland, work on mutualism in host-parasite relationship, 171

Coghill, functional differentiation of central nervous system, 84-5

Colpoda, 145-6

Compatible interaction in fungi, 134

Compatibility and noncompatibility, in tissue transplantation, 190-2; importance of stroma formation, 191

Compensatory hyperplasia, without functional overload, 200-1; in metanephros of embryonic chick, 201; in liver of parabiosed rats, 201; relation of hormones to, 201

Competence, 35

Complimentariness, 197, 204

Conalbumin, 11

Cystein, in roots, 104

Cytochrome oxidase, 82

Cytoplasm, interactions between nucleus and, 219-20

Determination, 34

Differentiation, and induction, 33-5; templates in, 46, 198; enzymatically active proteins in, 74-5, 79-80; cholinesterase in neural differentiation, 85-6; of cell strains, 198; relation of nucleus and cytoplasm to, 208; relation of nuclear variations to, 208-11; studied by nuclear transplantation, 212-19; in individual cells compared with groups of cells, 218

Dipeptidase, in plant growth, 97-105

Embryonic development, relation of cellular antigens to, 66-7; relation of enzymatically active proteins to, 73-4; development of enzymes during, 76; spatial

distribution of enzymes in, 76-7; nuclear transplantation between somatic cells and effects on, 212-19

Embryonic induction, 33-5; relation of evocation and individuation to, 36; mechanism of, 37; immunological techniques in a study of, 38-9; relation of antigens to, 45; importance of the reacting system in, 47; *see also* Induction

Embryonic transplants, 179; to anterior chamber of eye, 178-80; development of function in, 180; virus multiplication in, 180-1; susceptibility to carcinogenetic agents, 181-2

Embryo-specific proteins, 12-3

Endocrine glands, transplantation of, 178

Enzymes, spatial arrangement in embryonic induction, 48; as specific proteins in embryonic development, 73-4; as functional and structural proteins, 74-5; development of, 76; spatial distribution in embryo, 76-7; in muscle mitochondria, 82; in neural differentiation, 85-6; in plant growth, 93-117

Epigenesis, of red blood cell antigens, 5; of hemoglobin, 7; of blood plasma proteins, 8

Evocation, 35

Exudation, by microorganisms, 144-9; in presence of antisera, 144-6; results in loss of volume in paramecia, 144-5; by colpoda, 145; by tetrahymena, 145; by tapeworm, 146; by bacteria, 147; protein-carbohydrate character of, 148-9; by paramecia during conjugation, 149

Fertilization, in *Achlya,* 123

Fetuin, 13

Forsythia, incompatibility factors in, 126-9, 136

Fungi, incompatibility specificities in, 130-8

Genes, interaction and antigenic specificities of, 57; multiple antigenic specificities of, 61-6; *see also* Antigenic specificities

Glycine oxidase, in plant growth, 97-105

Greene, 177

Growth, inhibition by antibodies, 160-1; 165; specific control of, 198-200; self-limiting, 199; effects of homologous organ extracts on, 202-3

Growth, of plant cells, 93-117

Growth pattern, disturbance in microorganisms by antibodies, 149-51; alterations in bacteria, 149-50; alterations in tetrahymena, 150-1

Harrison, 141
Hemoglobins, epigenesis of, 7-8; autosynthetic, 7; embryonic and adult types, 7
Heterochromatin, 210
Heterologous transplantation, of embryonic tissues, 182-3; of human organs, 183; of cancer tissue, 185; the problems of compatibility and noncompatibility in, 190-2
Heteromorphic incompatibility in plants, 126
Heterosynthesis and early differentiation, 21-5; in *Drosophila*, 21; in mammals, 22-3; in amphibians and birds, 23
Heterosynthetic serum proteins, 13
Homologous compensation, as a growth phenomenon, 196; explanation of, 196; in metanephros of embryonic chick, 201; in liver of parabiosed rats, 201; *see also* Compensatory hyperplasia
Homologous growth effects, relation of templates and antitemplates to, 198-9; in metanephros of embryonic chick, 201; in liver of parabiosed rats, 201; after injection of cell debris, 202
Homologous organ extracts, effects on growth of cells in tissue culture, 202-3
Homologous organ-specific effects, 197, 199, 201, 204; after injection of cell debris, 202
Homologous transplantation, growth and differentiation in, 179; of embryonic tissues, 179-82; susceptibility to infection in, 180; susceptibility to carcinogenetic agents in, 181-2; of cancer tissue, 185
Homomorphic incompatibility in plants, 126-7
Hormones, fungal sex, 121-5; production in transplants, 178; in compensatory hyperplasia, 201
Host-parasite relationships, factors of parasite invasiveness, 157, 169; factors of host immunity, 157-8; mutualism in, 171
Hybridization, lethal effects of in amphibia, 219, 222
Hybrids, antigenic specificities in, 58-9; of pigeons and problems of antigens, 59-60; lethal character in amphibians, 221-2; utilization of chromatin precursors in amphibians, 223
Hyperplasia, 200-1
Hyperplastic and precancerous tissues, 183-4; transplantation of, 184

Immobilization reaction, in microorganisms, 141-3; caused by antibodies, 142; relation of cilia to, 142; importance in organization of paramecia, 143; compared with agglutinative, precipitative, complement-fixing tests, 143; specificity in bacteria and protozoa, 144
Immunity, innate and acquired, 158; antibodies and acquired immunity, 158-9; innate immunity and genetic control, 159; relation to rate of reproduction, 160-1; 168; acquired immunity superimposed upon innate immunity, 163; *see also* Antibodies, Antigens, Antiserum
Immunological reversal of organ-specific compounds, 197-8
Incompatibility, in plants, 126-7; in heterologous transplantation in animals, 190-2; *see also* Transplantation
Incompatibility factors, 119-20, 124-39
Incompatibility specificities in higher fungi, 130-8
Individuation, 35
Induction, 33-5; qualitatively different substances in, 37; specific proteins in, 38; relation of antigens to, 45; importance of reacting system in, 47; *see also* Embryonic induction
Inductors, primary, 34; higher level, 34; in dorsal lip of blastopore, 35; cause cytolysis, 37
Inhibitors, of plant growth, 96, 105, 113-4
Innate immunity, in mammals, 159; in birds, 159; in mosquito, 160; effects of drugs on, 160; relation to parasiticidal substances, 160; relation to acquired immunity, 163; *see also* Immunity
Invertase, in plant growth, 97-105, 111-3
Irwin, 55
Isoagglutinins, 9; epigenetic character of, 10; *see also* Agglutination
Isotopes, use of in tagging antibodies, 203-4

King, 207

Lens, development of, 42; proteins of, 42-3; antisera to, 42-3, 203; regeneration of, 43
Lens proteins, 42; isolation of, 43; in lens regeneration from iris, 44
Lethal hybrids, in amphibia, 219-23; *see also* Nucleocytoplasmic interactions
Lewis, 129
Limb regeneration blastema, 44; reactions to antisera, 44-5
Linum, incompatibility factors in, 129, 139
Localization of parasites, mechanism of, 170-1

Macromolecular transfer, significance of, 4-5; in birds and amphibians, 16-8; in

Macromolecular transfer (*continued*)
rabbit blastocyst, 18-20; in mammals, 18-21, 23-4; in placenta, 19-20
Marrack, hypothesis of serologic agglutination, 153-4
Mating specificities and sexual hormones in plants, 121-4
Microsomes, relation of adenosinetriphosphatase to, 79, 81
Mitochondria, enzymes associated with, 82-3; of muscle and their enzymes, 82; in histogenesis of muscle, 84
Mitotic index, 200, 202
Moewus, 127-8
Multiple antigenic specificities of genes, 61-6; in blood of chickens, 62; in blood of cattle, 62-4
Mutualism, in host-parasite relationship, 171-2; *see also* Host-parasite relationships
Myosin, 44; adenosinetriphosphatase activity of, 75, 79-81

Neural differentiation, cholinesterase in, 84-9
Newcastle virus, inoculation of chick embryos with, 14; immunization of birds against, 16-7
Ninhydrin reacting substances, in roots of plants, 104
Noncompatible interactions, in fungi, 138; *see also* Incompatibility
Noncompatibility, in transplantation, 190-2; *see also* Transplantation
Nonnucleated ova, developmental potencies of, 214-5
Nuclear transplantation, 212-9; in unicellular organisms, 213-4; in *Acetabularia*, 213; in amoeba, 214; from embryonic cells to enucleated ova, 214; techniques of in amphibians, 216; from blastula and gastrula cells to enucleated ova, 217, 223; in hybrids, 219-23; xenoplastic, 223-4
Nucleocytoplasmic interactions, in nuclear transplantation, 212-9; in lethal hybrids of amphibians, 219-23
Nucleus, variation in somatic cells, 208-11; polyploidy in somatic cells, 210-1; biochemical variations of, 211; significance for cell differentiation, 211-2; transplantation between cells, 212-9; in amphibian hybrids, 219-21; *see also* Nuclear transplantation

Oenothera, 129
Organizer, 35-6
Ovalbumin, 11

Papazian, 132-3, 137
Paramecium, immobilization reaction in different strains, 143-4; antigenic change during conjugation of, 149
Parasite invasiveness and virulence, factors associated with, 169; role of specific enzymes in, 170; relation to elective localization of parasites, 170-1
Parasites, innate immunity to, 158-60; acquired immunity to, 165-7; invasiveness and virulence, 169; elective localization of, 170-1; mutualism in relation to host, 171-2
Parasiticidal substances, 160, 168
Phagocytosis, by various cells of chick embryo, 15; by chorio-allantoic membrane, 15-6
Phosphoproteins, in yolk of hen's egg, 11
Pigmentation, embryonic development of, 40, 47-8
Plant cells, growth of, 93-117
Plant growth, enzymes in, 93-117; proteins in, 93-117; glycine oxidase in, 97-105; proteinase in, 97-105, 115-7; dipeptidase in, 97-105; acid phosphatase in, 97-105, 111; invertase in, 97-105, 111-3; inhibitors of, 96, 104, 113-4
Pollen development and specific inhibitors, physiological differentiation, 124-9
Pollen germination, inhibition of, 124-30
Polyploidy, 210-11
Precancerous tissue, transplantation of, 183-4
Proteinase, in plant growth, 97-105, 115-7
Proteinogenesis, as a chemical basis of morphogenesis, 74-5
Proteins, in amphibian serum, 10-11; in bird serum, 16; in mammalian serum, 18; of lens, 42-3; enzymatically active proteins in differentiation and morphogenesis, 74-5; catalytic activity of, 75; in plant growth, 93-117; *see also* Serum proteins

Quercetin, 127-8

Raper, 119
Reacting system, importance of in induction, 35
Red blood cell antigens, epigenesis of, 5-7; relation to genic factors, 7; *see also* Antigens
Respiratory enzymes, in muscle mitochondria, 82-4
Robinson, 93
Roots, water content of, 99-100; arginine in, 104; ninhydrin reacting substances in, 104; cystein in, 104

Rous chicken sarcoma, 184-5, 186
Rutin, 127-8

Schechtman, 3
Schizophyllum, 130-9
Secretion of water, in plants, 95
Serologic agglutination, mechanism of, 152-3; spatial proximity as a factor in, 153; Marrack hypothesis of, 153-4
Serum proteins, in developing chick embryo, 10-11; autosynthetic and heterosynthetic, 13; transfer in birds and amphibians, 16-8; permeability of rabbit blastocyst to, 18-9; transfer in mammals, 18-21; permeability of placenta to, 19-20; in embryo compared with adult, 21
Sex hormones, and mating specificities in plants, 121-4; in *Thraustotheca,* 124-5
Shen, 73
Shope papilloma, 178-9, 181, 183, 186, 192
Skin transplantation, 47-8; *see also* Transplantation
Somatic nuclei, variations among, 208-11; giant, 210; polyploidy in, 210-1; biochemical variations of, 211; significance in cell differentiation, 211-2; *see also* Nucleus, and Nuclear transplantation
Sparing phenomenon, 163
Specific growth control, concept of, 198-200
Spemann, work on induction, 33-4
Stroma, formation and importance of in transplants, 188-9; connective tissue as an inducer of, 189
Succinic dehydrogenase, 82
Succinoxidase, in embryonic brain, 86, 88
Sugar, affect on growth in plants, 95, 105-13

Taliaferro, 157
Templates, in differentiation, 46, 198; relation to antitemplates, 199-200
Tetrahymena, 145, 150

Thraustotheca, sex hormones in, 124-5
Tissue culture, homologous organ extracts in, 202-3
Transplantation, of skin, 47-8; of fetal tissue in cattle twins, 67; of skin in identical and non-identical twins, 68; of adult tissues to anterior chamber of eye, 178; of endocrine glands, 178; of embryonic tissues, 179-83; of tissues to brain, 181; of hyperplastic and precancerous tissues, 183-4; of cancer tissue, 185-7; importance of site of, 187-9; stroma formation in, 188-9; compatibility and noncompatibility in, 190-2; *see also* Nuclear transplantation
Transplantation site, 187-90; importance of stroma formation at, 188-9
Transplants, autologous, homologous, heterologous, 177, 187; normal functioning in anterior chamber, 178; endocrine activity of, 178; of embryonic tissues, 179; to brain, 181; susceptibility to carcinogenetic agents, 181-2; of hyperplastic and precancerous tissue, 183-4; of cancer tissue, 185-7; stroma formation in, 188-9; problems of compatibility and noncompatibility in, 190-2
Treponema, 143
Tumors, 183; virus induced, 186-7; importance of transplantation site, 188-9; of brain and their transplantability, 189

Virus, growth in chick embryo, 14; multiplication in anterior chamber transplants, 180
Virus induced tumors, 186-7

Water content, of bean root, 99-100
Water-molds, 121-5
Weiss, 195
Woerdeman, 33

Yolk, phosphoproteins in, 11; comparison of between amphibian and bird, 17